ADHESIVES TECHNOLOGY COMPENDIUM 2019

adhesion ADHESIVES & SEALANTS
Industrieverband Klebstoffe e. V.
(German Adhesives Association)

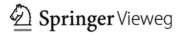

Editor: adhesion ADHESIVES & SEALANTS
Abraham-Lincoln-Straße 46
D-65189 Wiesbaden
Phone +49 (0) 6 11-78 78-2 83
www.adhaesion.com
Email: adhaesion@springer.com

Supported by:
Industrieverband Klebstoffe e. V.
German Adhesives Association
Völklinger Straße 4 (RWI-Haus)
D-40219 Düsseldorf
Phone +49 (0) 2 11-6 79 31 10
Fax +49 (0) 2 11-6 79 31 33

Publishing house: Springer Vieweg | Springer Fachmedien Wiesbaden GmbH
Abraham-Lincoln-Straße 46
D-65189 Wiesbaden
www.springer-vieweg.de

CHEMISCHE INDUSTRIE
Berufsgruppe Bauklebstoffe

Fachverband Klebstoff-Industrie Schweiz

Industrieverband
Klebstoffe e.V.

vereniging lijmen en kitten

| Austria (A) | Swiss (CH) | Germany (D) | Netherland (NL) |

All company profiles and sources are based on the information provided by the respective companies (as at July 2019). The publishing house does not accept any responsibility for the completeness and correctness of the information.

Layout: satzwerk mediengestaltung · D-63303 Dreieich

ISBN 978-3-658-27072-8

Nominal sum: € 25.90

Dear Reader,

Today, there is hardly any industry or business sector that does not rely on the use of adhesive bonding, which is an innovative, reliable joining method. It is indispensable when it comes to joining different materials while preserving their properties and offering long-term stability. The opportunities for using new, reliable construction methods can only be exploited in conjunction with innovative adhesive systems. In addition to the joining process itself, other functionality can also be integrated into bonded components, for instance by balancing the differing dynamics of joined parts, providing corrosion protection or vibration absorption or sealing against liquids and gases. More than any other joining technology, adhesive bonding allows sophisticated designs to be created, because it offers the best possible combination of technological sophistication, cost-effectiveness and a low environmental impact. Adhesive bonding is without a doubt the key technology of the 21st century.

The German adhesives industry is particularly important in this respect. It is a technology leader on both European and global markets. The demand from other countries for adhesives and sealants that are made in Germany remains high. The German adhesives industry exports more than 40 % of its products every year. In addition, the foreign subsidiaries of German adhesives manufacturers generate sales of more than 8 billion, which gives the German adhesives industry a 20 % share of the global market with adhesives and sealants designed in Germany.

Alongside its leading technological position, the German adhesives industry is also highly important in economic terms, which is something that the general public is almost entirely unaware of. Every year more than 1.5 million tonnes of adhesives and sealants and more than 1 billion square metres of adhesive films and tapes are manufactured in Germany, which results in total industry sales of almost 4 billion. The use of these adhesive systems generates potential added value of more than 400 billion. This huge amount corresponds to around 50 % of the contribution made by manufacturing industry and the construction sector to the German gross domestic product (GDP). In other words, around 50 % of the goods produced in Germany have a connection with adhesives.

The German Adhesives Association (IVK) represents the technical and economic interests of 147 manufacturers of adhesives, sealants, raw materials and adhesive tapes, key system partners and scientific institutes. It is the world's largest and, in terms of its comprehensive portfolio of services, the world's leading association for adhesive bonding technology.

In this compendium the German Adhesives Association, in cooperation with its sister associations – the Association of the Adhesives Industry in Switzerland (FKS), the Association of the Austrian Chemical Industry (FCIO) and the Association of the Dutch Adhesives and Sealants Industry (VLK) – gives an insight into the world of the adhesives industry.

One of the main tasks of these associations is to provide regular information about adhesive bonding as a key technology, as well as about manufacturers of innovative adhesive systems and the activities of the industry's organisations. This compendium contains important facts about the adhesives industry and its associations, as well as describing the extensive product and service profiles of adhesives manufacturers, key system partners and scientific institutes.

Together with the editorial team of "adhäsion KLEBEN & DICHTEN", we are pleased to present the 6th edition of our Adhesive Technology Compendium.

Dr. Boris Tasche
President of
Industrieverband Klebstoffe e. V.

Ansgar van Halteren
Senior Executive of
Industrieverband Klebstoffe e. V.

German
Adhesives
Association
Industrieverband Klebstoffe e.V.

adhesion ADHESIVES
SEALANTS

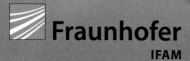
Fraunhofer
IFAM

EMICODE®
GEV

Industrieverband
Klebstoffe e.V.

**Company Profiles
Equipment and Plant Manufacturing**

**Company Profiles
Adhesive Technology Consultancy
Companies**

List of Advertisers

Cover: Auftrag eines Gap-Fillers (© bdtronic)

Structural Adhesives

Requirements for Modern Carbody Adhesives

A new generation of SikaPower structural adhesives is based on a modular system that makes it possible to combine a wide range of material properties. The basis is a special low-viscosity epoxy resin matrix. The innovative feature of this system is its combination of low viscosity and outstanding wash-out resistance. These adhesives also cover a very large bonding range. To be specific, they exhibit excellent properties on the zinc-magnesium coated steel substrates we have been hearing so much about lately.

Bonding as a joining method is now being discussed particularly often in connection with the implementation of lightweight construction concepts. For example, special adhesive systems have been developed for the bonding of mixed structures (e.g. steel/aluminum or steel/CRP) so that the bond remains intact during heat-curing even when the individual substrates have different thermal expansion coefficients. Adhesive systems such as these can make a significant contribution to the construction of lightweight structures. These adhesives are marketed by Sika under the name SikaPower MBX.

When discussing new, innovative lightweight construction concepts and the appropriate joining methods, pure steel/steel or aluminum/aluminum bonds almost take a back seat. We should remember, however, that most applications of structural adhesives still involve such conventional adhesions. Steel is the world's most important material for the construction of automotive structures, followed by aluminum. It is therefore essential to regularly reassess the properties of conventional structural adhesives.

Wash-out resistance

In the past, structural adhesives were often highly viscous. After hot application, these types of systems cool down very quickly on the cold sheet metal and, like hot-melt adhesives, rebuild the high initial viscosity, which leads to very high washout resistance. Fully heatable application equipment is required for the processing of these adhesives. A new adhesive technology now combines low viscosity (Figure 1) with outstanding wash-out resistance. This makes it possible to pump and convey the adhesives at lower temperatures. Consequently, the cost of procuring and maintaining the systems and the cost of energy for operating them can be decreased without sacrificing the usual features. For the results of a laboratory test for evaluating wash-out resistance shown in Figure 2, an adhesive bead was applied to a strip of sheet metal, half of which was then immersed in water at a defined speed and rotated at a specific bath temperature. Then, the deformation of the adhesive bead was assessed in comparison with the unloaded upper part (Figure 2). The lower viscosity also improves the compression and thus leads to a more stable joining process.

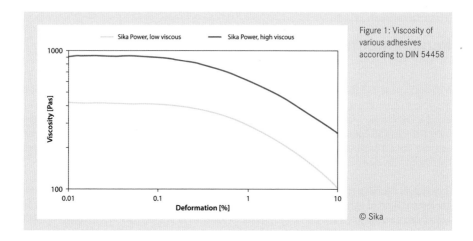

Figure 1: Viscosity of various adhesives according to DIN 54458

© Sika

Application

The lower viscosities of these adhesive systems lead to lower application pressures (Figure 3). With nozzle diameters of 0.8 1.4 mm, which are often used in bodywork, the low-viscosity system achieves 30–40% lower pressures, depending on the discharge rate. An application temperature of at least 45°C at the nozzle is gradually obtained across the feed zones to ensure consistent application conditions throughout the year (Figure 4). If the discharge temperature at the nozzle is increased to 55 °C, thin-jet application with a component distance of up to 15 mm is possible using a nozzle with a diameter smaller than 0.8 mm. The materials can be applied using either the E-Swirl (Figure 5) or the Air-Swirl method. Low material viscosity offers even more advantages with respect to application.

Figure 2: Wash-out resistance of various adhesives SikaPower high viscous (left), SikaPower low viscous (right)

© Sika

- For one, the stringiness at the end of the adhesive bead is reduced, which reduces smudges on the component surfaces and on the tools. The lower pressures lead to longer service lives for individual components such as seals. This increases the availability of the system and reduces the maintenance costs.

- Additionally, when retracting the downstream container, the pump can be vented directly without the need for preheating time. This simplifies the logistics chain for materials procurement and increases the operational safety.

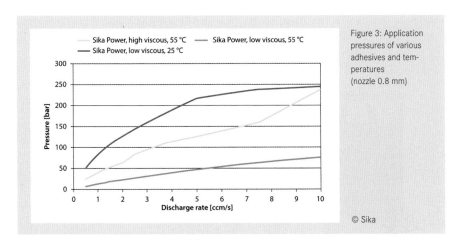

Figure 3: Application pressures of various adhesives and temperatures (nozzle 0.8 mm)

© Sika

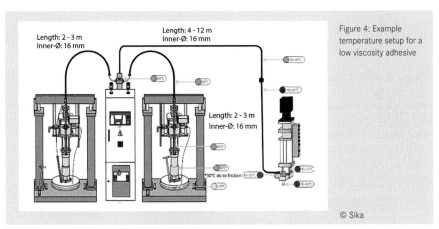

Figure 4: Example temperature setup for a low viscosity adhesive

© Sika

- The lower pressures lead to longer service lives for individual components such as gaskets. This increases the availability of the system and reduces the maintenance costs.

- Additionally, when a new container is placed under the flower-plate the pump can be vented directly without pre-heating time. This simplifies the logistics chain for materials procurement and increases the operational safety. The lower viscosity also leads to significant advantages in the process. Components can still be joint with no problems, even after prolonged breaks following the application of the adhesive, as the viscosity increases only slightly after cooling.

- The wetting on the oiled component surface is also better, which means that adhesive beads remain in place even during rapid robot movements.

- Adhesive leaks, e.g. in hem flange bonding, can also be removed more easily. This leads to significant time savings for the process.

Figure 5: E-Swirl Application with SikaPower low viscosity

© Sika

New substrates

Electrolytically and hot-dip galvanized substrates are still important in steel-intensive body-work. The proportion of hot-dip galvanized substrates has increased significantly in recent years though. The steel industry has been offering new grades based on hot-dip galvanized substrates for quite some time. Two trends are now emerging. One of these is a trend towards

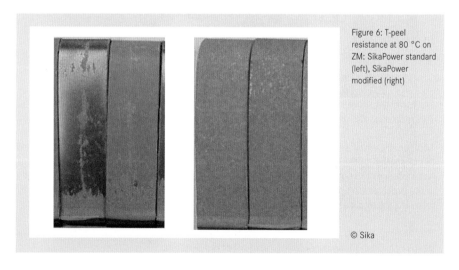

Figure 6: T-peel resistance at 80 °C on ZM: SikaPower standard (left), SikaPower modified (right)

© Sika

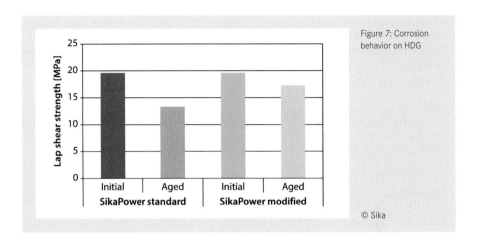

Figure 7: Corrosion behavior on HDG

© Sika

zinc-magnesium surfaces (ZM). In addition to zinc and aluminum, these coatings also contain magnesium in the melt. The advantage of materials such as these lies in the improved corrosion properties as well as improved tribological behavior. Until now, these substrates show reduced adhesion when bonded with common structural adhesives, which was particularly evident in the peel test at 80 °C. The adhesives can be modified to specifically improve adhesion and corrosion resistance by using specially developed components (Figure 6). This recognizably increases the peel strength. Another trend is the application of inorganic substances to the surface in order to improve tribological behavior during forming. Sulfur, phosphorus,

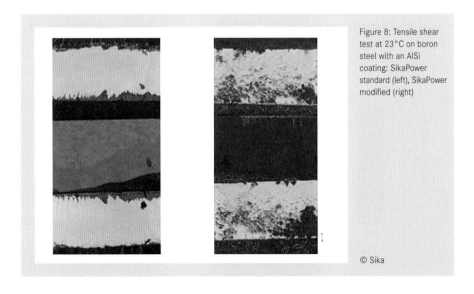

Figure 8: Tensile shear test at 23 °C on boron steel with an AlSi coating: SikaPower standard (left), SikaPower modified (right)

© Sika

calcium and potassium compounds play an important role in this. It is imperative that the corrosion behavior of substrates such as these is tested before they are used for the bond. There are clear differences between the various qualities available on the market. In this case as well, particularly on hot-dip galvanized steel (HDG), the above-mentioned modification of the adhesive can lead to an improvement in the corrosion behavior (Figure 7).

Another example of new substrates are high-strength boron steels, which are given an aluminum-silicon coating (AlSi) to improve their corrosion resistance. As these coatings often adhere poorly to the steel and because of their brittleness, bonding with structural adhesives leads to delamination of the AlSi coating. Innovative structural adhesives can prevent this delamination (Figure 8).

Aluminum

It is a known fact that passivation is necessary for the stable, corrosion-resistant bonding of aluminum. Very good results are produced with TiZr conversion layers, for example. Figure 9 shows a comparison with the same adhesive on passivated and unpassivated aluminum before and after ageing, in this case 10 cycles of VDA alternating corrosion test. The advantage of a passivation such as this is clearly demonstrated. Nevertheless, it is necessary to adapt adhesives specifically for use on aluminum.

Mechanical properties

For structural adhesives, it is of course important that the mechanical properties are suitable, even if it seems that an optimum has been achieved here. The primary task of such materials is to transfer high forces and to increase the stiffness and the fatigue strength. Another important mechanical property is the crash resistance of structural adhesives. With crash-opti-

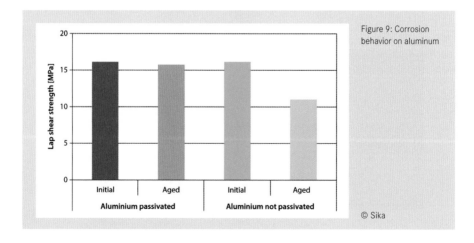

Figure 9: Corrosion behavior on aluminum

© Sika

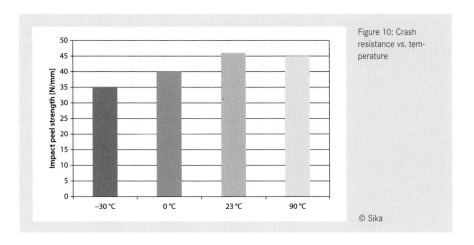

Figure 10: Crash resistance vs. temperature

© Sika

mized adhesives, the adhesive seam does not burst under sudden load, but allows the component to deform without failure, resulting in significantly better energy absorption in the component. In order to achieve these properties, it is necessary to combine the brittle epoxy resin with a second polymer, a toughening agent. The fact that Sika has developed its own toughening technology, and also produces its own polymers, makes it possible to design the polymers so that they are optimally adapted to the desired material properties. The degree of crash resistance desired can be set according to the requirements. With particularly highly crash-resistant adhesives, it is important that this outstanding property can also be maintained over the entire temperature range of 40 to 80 °C (Figure 10). Also these materials can be processed very reliably due to the very large range of baking temperatures and duration.

Additional properties

The storage stability is usually at least 10 months at 25 °C. To differentiate between the various adhesives in a body, it is possible to color the adhesives differently. The contrast enhancement for image-drive inspection systems can be optimized with targeted coloring. The material can also be mixed with glass beads of various diameters without changing the adhesive properties. The glass beads act as spacers, e.g. in the hem flange bonding, and thus ensure uniform mechanical properties in the component. The innovative modular system of the latest generation of SikaPower adhesives allows the above-mentioned properties to be specifically combined and adapted to customer-specific requirements. So it stands to reason that with the described adhesive system, Sika is making a significant contribution to stabilizing automobile manufacturing processes.

*Michael Gutgsell (gutgsell.michael@ch.sika.com)
is head of the "Body Shop Adhesives & Sealants"
department at Sika Technology AG in Zurich.*

COMPANY PROFILES

Adhesive Producer
Raw Material Supplier

3M Deutschland GmbH

Carl-Schurz-Straße 1
D-41453 Neuss
Phone +49 (0) 21 31-14 33 30
Fax +49 (0) 21 31-14 32 00
E-Mail: kleben.de@mmm.com
www.3M-klebtechnik.de

Member of IVK

Company

Year of formation
1951

Size of workforce in Germany
6,700

3M Deutschland GmbH
Headquarter – Neuss
Manufacturing sites – Hilden, Kamen
European Distribution Center – Jüchen
3M Oral Care Solutions Division – Seefeld,
Landsberg/Lech
3M Separation and Purification Sciences Division –
Wuppertal
3M Health Information Systems (HIS) – Berlin

Subsidiaries 3M Deutschland GmbH
Dyneon GmbH – Burgkirchen
TOP-Service for Lingualtechnik GmbH – Bad Essen
Wendt GmbH – Meerbusch, Niederstetten

Further 3M Companies in Germany
3M Medica – Neuss
3M Technical Ceramics – Kempten

Sales channels
Distributor + direct

Management
Christiane Grün, Manfred Hinz, Oliver Leick

Range of Products

1-part structural adhesives
2-part structural adhesives

Types of adhesives
Hot melt adhesives
Reactive adhesives
Dispersion adhesives
Solvent-based adhesives
Spray adhesives
Pressure-sensitive adhesives

Types of sealants
Acrylic sealants
Butyl sealants
PUR sealants
SMP/STP sealants

Raw materials
Fillers
Polymers

Equipment, plant and components
for conveying, mixing, metering and for adhesive
application
for surface pretreatment
for adhesive curing
adhesive curing and drying
measuring and testing

For applications in the field of
Paper/packaging
Bookbinding/graphic design
Wood/furniture industry
Construction industry, including floors, walls and
ceilings
Electronics
Mechanical enineering and equipment construction
Automotive industry, aviation industry
Textile industry
Adhesive tapes, labels
Hygiene
Household, recreation and office

ADHESIVE TECHNOLOGIES

Adtracon GmbH
Hofstraße 64
D-40723 Hilden
Germany
Phone +49 (0) 21 03- 253 1710
Fax +49 (0) 21 03- 253 1719
Email: info@adtracon.de
www.adtracon.de

Member of IVK

Company

Year of formation
2002

Size of workforce
15

Ownership structure
Dr. Roland Heider, ISB, KfW

Sales channels
Direct and distributors

Contact partners
Management:
Dr. Roland Heider

Further information
Adtracon specializes in the development,
production and marketing of reactive hot
melt adhesives. We also offer technical
and laboratory services and consultancy
services.

Range of Products

Types of adhesives
Reactive hot melt adhesives

For applications in the field of
Bookbinding/graphic design
Woodworking/furniture industry
Automotive industry, aviation industry
Textile/filter/shoe and leather industry

Alberdingk Boley GmbH
Düsseldorfer Straße 53
47829 Krefeld, Germany
Phone +49 (0) 21 51-5 28-0
Fax +49 (0) 21 51-57 36 43
Email: info@alberdingk-boley.de
www.alberdingk-boley.de

Member of IVK

Company

Year of formation
1828

Size of workforce
450 (worldwide)

Ownership structure
family-owned

Subsidiaries
Alberdingk Boley Leuna GmbH, Leuna
Alberdingk Boley, Inc., Greensboro, USA,
Alberdingk Resins (Shenzhen) Co. Ltd.,
Shenzhen, China,
Thai Castor Oil Industries Co., Ltd.,
Bangkok, Thailand

Sales channels
worldwide

Contact partners
Management:
Timm Wiegmann
Sales and Marketing

Application technology and sales:
Irene Tournee,
Technical Marketing Adhesives

Marketing Coatings:
Johannes Leibl, Manager Sales Dispersions

Range of Products

Raw materials
Polymers:
Polyurethane dispersions
Acrylate dispersions
Styrene acrylic dispersions
Vinylacetate copolymer dispersions
UV-curable dispersions

For applications in the field of
Adhesives:
Tapes, Labels,
Paper/Packaging
Construction
Automotive
Foils/flexible packaging

Coatings:
Wood/Furniture
Metal/Plastics
Construction including floors, walls and
ceilings
Textile/Leather
Film coatings
Primer

ALFA Klebstoffe AG
Vor Eiche 10
CH-8197 Rafz
Phone +41 43 433 30 30
Fax +41 43 433 30 33
Email: info@alfa.swiss
www.alfa.swiss

Member of FKS

Company

Year of formation
1972

Size of workforce
63

Ownership structure
Family Private Limited (AG) Company

Subsidiaries
ALFA Adhesives, Inc. (Partner)
SIMALFA China Co. Ltd.

Sales channels
International distribution network

Contact partners
Management:
info@alfa.swiss

Application technology and sales:
info@alfa.swiss

Further information
ALFA Klebstoffe AG, an innovative family
business, focused on the development,
production and distribution of water-based
adhesives and hotmelts; therefore, ALFA
Klebstoffe AG offers a significant benefit
regarding the economic and ecologic design
of the gluing process of their customers.

Range of Products

Types of adhesives
Dispersion adhesives
Hot melt adhesives
Pressure-sensitive adhesives

For applications in the field of
Foam converting industry
Matresses production
Upholstery
Paper/packaging
Bookbinding/graphic design
Wood/furniture industry
Automotive industry, aviation industry
Hygiene

APM Technica AG

Max-Schmidheinystrasse 201
CH-9435 Heerbrugg
Phone +41 (0) 71 788 31 00
Fax +41 (0) 71 788 31 10
Email: info@apm.technica.com
www.apm-technica.com

Member of FKS

Company

Year of formation
2002

Size of workforce
140

Ownership structure
Private Shareholder

Subsidiaries
APM Technica Philippines, APM Technica
GmbH, Abatech, Polyscience AG

Sales channels
Direct & Distributor

Contact partners
Management:
Andreas Hedinger, Giorgio Rossi

Application technology and sales:
APM Technica AG is the full-service provider
in the field of adhesives and surface technology.

Further information
www.apm-technica.com

Range of Products

Types of adhesives
Reactive adhesives

Types of sealants
Polysulfide sealants
Silicone sealants
MS/SMP sealants

Equipment, Plant and Components
for surface pretreatment
for adhesive curing
adhesive curing and drying

For applications in the field of
Electronics
Mechanical engineering and equipment
construction
Automotive industry, aviation industry
Medical and optic industry

ARDEX GmbH

Friedrich-Ebert-Straße 45
D-58453 Witten
Phone +49 (0) 2 30 26 64-0
Fax +49 (0) 2 30 26 64-3 75
Email: info@ardex.de
www.ardex.de

Member of IVK, FCIO, VLK

Company

Year of formation
1949

Size of workforce
3,000

Ownership structure
Private

Subsidiaries
Australia, Austria, Bulgaria, China, Czech Republic, Denmark, Finland, France, Germany, Hungary, Hong Kong, India, Ireland, Italy, Korea, Luxembourg, Mexico, New Zealand, Norway, Poland, Romania, Russia, Singapore, Spain, Sweden, Switzerland, Taiwan, Turkey, UK, United Arab Emirates, USA

Contact partners
Management:
Mark Eslamlooy (CEO)
Dr. Ulrich Dahlhoff
Dr. Hubert Motzet
Uwe Stockhausen
D. Markus Stolper

Application technology and sales:
Daniel Händle

Range of Products

Types of adhesives
Dispersion adhesives
Pressure-sensitive adhesives

Types of sealants
Acrylic sealants
Butyl sealants
PUR sealants
Silicone sealants
MS/SMP sealants

For applications in the field of
Construction industry, including floors, walls and ceilings

Performance Elastomers

ARLANXEO Deutschland GmbH
Chempark Dormagen, Building F41
Alte Heerstraße 2
D-41540 Dormagen
www.arlanxeo.com

Member of IVK

Company

ARLANXEO is a world-leading synthetic rubber company with sales of around EUR 3.2 billion in 2018, about 3,900 employees and a presence at 20 production sites in nine countries. ARLANXEO was established in April 2016 as a joint venture of LANXESS and Saudi Aramco. On January 1st, 2019, Saudi Aramco became the sole owner of ARLANXEO.

Headquartered in Maastricht, Netherlands, the company's core business is the development, manufacturing and marketing of synthetic high-performance rubber.

Contact persons
For Baypren® and Levamelt®
Dr. Martin Schneider
High Performance Elastomers
Phone: +49 221 8885 5908
E-Mail: martin.schneider@arlanxeo.com

For X_Butyl® RB
Dr. Thomas Rünzi
Tire & Specialty Rubbers
Phone: +49 221 8885 4829
E-Mail: thomas.ruenzi@arlanxeo.com

Range of Products

Raw materials for the production of adhesives and sealants

Special grades from the product lines Baypren® (chloroprene rubber), Levamelt® (ethylene-vinyl acetate copolymers) and X_Butyl® (butyl rubber) are tailor-made for the use in adhesive applications. The synthetic rubbers are offering unique properties regarding elasticity, polarity and tackiness, making them particular suitable for versatile applications in the adhesive and sealant industry:

- Baypren®: First choice for solvent-borne contact adhesive
- Levamelt®: Base polymer for pressure sensitive adhesives and modifier for structural adhesives and hot melts
- X_Butyl®: Basis for pressure sensitive adhesive and sealants

artimelt AG

Wassermatte 1
CH-6210 Sursee
Phone +41 41 926 05 00
Email: info@artimelt.com
www.artimelt.com

Member of FKS

Company

Year of formation
2016

Ownership structure
artimelt is a family owned company in Switzerland.

Subsidiary, Representative
artimelt Inc., Tucker, GA 30084, USA
Formosa Nawonsuith Corp. (FNC), 11552 Taipei, Taiwan

Sales channels
Direct sales and agents

Managing director
Walter Stampfli

Contact partner
Christoph Lang
Phone: +41 41 926 05 28
Email: christoph.lang@artimelt.com

Further information
artimelt combines many years of experience with a high degree of innovation; for future-oriented and sustainable solutions. artimelt is always looking for something special and supports its customers on a partnership basis to them be competitive in their market.

Range of Products

Types of adhesives
Hot melt adhesives

For applications in the field of
Medical products
Labels
Tapes
Packaging
Security systems
Building
Specialities

ATP adhesive systems AG
Sihleggstrasse 23
CH-8832 Wollerau
Phone +41 (0) 43 888 15 15
Fax +41 (0) 43 888 15 10
Email: info@atp-ag.com
www.atp-ag.com

Company

Year of formation
1989

Size of workforce
250

Managing partners
Managing Director:
Daniel Heini

Ownership structure
Privately owned

Sales channels
Direct sales channels
Graphic Distributors

Further information
ATP is a leading manufacturer of high quality technical tape solutions for the automotive, foam, graphic, label, semi-structural composite, building and construction industries. With our extensive technical and marketing knowledge, our passion for developing customer-focused solutions and our committed employees that go that little bit further, success with ATP is a given.
ATP is producing high quality single- and double-sided adhesive tapes on very modern coating machines in Germany since 1991. Using a broad range of adhesive and support materials, ATP produces single- and double-sided pressure sensitive adhesive tapes, transfer tapes and heat-seal films. The adhesive formulations are solvent free and are exclusively developed by ATP.

Range of Products

Types of adhesives
Dispersion adhesives
Pressure-sensitive adhesives
Self adhesive tapes
(Single- and double-sided tapes with different carriers)

For applications in the field of
Paper/packaging
Bookbinding/graphic design
Wood/furniture industry, including floors, walls and ceilings
Electronics
Automotive industry, aviation industry
Textile industry
Adhesive tapes, labels

ATP's production methods meet the most modern technological requirements. The products are developed and manufactured under a quality management system which is certified according to ISO 9001, ISO 14001, ISO 50001 and ISO/TS 16949.
ATP always strives to exceed customer expectations by developing customer aligned solutions which offer technical advantages, competitively and quickly.

Avebe U. A.

Prins Hendrikplein 20
NL-9641 GK Veendam
Phone +31 (0) 598 66 91 11
Fax +31 (0) 598 66 43 68
Email: info@avebe.com
www.avebe.com

Member of VLK

Company

Range of Products

Established in
1919

Employees
1,311

Ownership structure
Cooperation of farmers

Sales channels
Direct and through specialized
distributors worldwide

Further information
Avebe U. A. is an international Dutch starch
manufacturer located in the Netherlands and
produces starch products based on potato
starch and potato protein for use in food,
animal feed, paper, construction, textiles and
adhesives.

Raw materials
Dextrins and Starch based adhesives
Starch ethers

For applications in the field of
Paper and packaging
Paper sack adhesive
Tube winding adhesive
Remoistable envelope adhesive
Protective colloid in polyvinyl acetate
based dispersions
Water activated gummed tape adhesive
Wallpaper and bill posting adhesive
Additive – Rheology modifier for cement
and gypsum based mortars and tile
adhesives
Water purification

BASF SE
D-67056 Ludwigshafen
Phone +49 (0) 6 21-60-0
Email: industrial-adhesives@basf.com
pressure-sensitive-adhesives@basf.com
info-pib@basf.com
www.basf.com

Member of IVK, VLK

Company

Year of formation
1865

Size of workforce
approximately 122,000 employees
(as of year end 2018)

Range of Products for Adhesives
1. Acrylic dispersions (water-based)
2. Acrylic hot melts (UV-curable)
3. Polyurethane dispersions (PUD)
4. Styrene-butadiene dispersions
5. Polyisobutene (PIB)
6. Polyvinylpyrrolidone (PVP)
7. Polyvinylether (PVE)
8. Auxiliaries:
   • Crosslinking agents
   • Defoamer
   • Thickener
   • Wetting agents
9. Additives
   • Antioxidants
   • Light stabilizer
   • Photoinitiators/Curing agents
   • Others

Email contacts
1 – 9: pressure-sensitive-adhesives@basf.
   com; industrial-adhesives@basf.com
5: info-pib@basf.com

Range of Products

Comments
BASF is one of the world's leading manu-facturers of raw materials and additives for adhesives.

With its technologies, BASF offers its cus-tomers an alternative to traditional fastening methods. The additive portfolio helps to improve products no matter which adhesive technology is used. BASF's polymer disper-sions, UV curable hot melts and auxiliaries are an excellent solution for the production of high-quality self-adhesive products such as labels, tapes and films.

For high sophisticated adhesives, BASF assists with know-how, reliability and safety based on decades of experience. BASF constantly develops and tests new products – according to customer requirements.

BASF also supplies a wide range of polyiso-butenes (PIB) with low-, medium- and high molecular weight. Beyond other applications, Glissopal® (LM PIB) is used as tackifier to adjust stickiness of adhesive formulations. The applications for Oppanol® (MM and HM PIB) range from adhesives, sealants through to chewing gum base.

BCD Chemie GmbH
Schellerdamm 16
D-21079 Hamburg
Phone: +49 40 77173-0
Fax: +49 40 77173 2640
Email: info@bcd-chemie.de
www.bcd-chemie.de

Member of IVK

Company

Size of workforce
approx. 145 employees

Managing partners
Ronald Bolte (Chairman)
Lothar Steuer

Range of Products

Raw materials
Additives: Dispersing agents, defoamer, adhesion promoter, hydrophobing agents, wetting agents, thickener, rheological additives, silanes, flame retardant agents

Resins: Thermoplastic acrylates, styrene acrylic and pure acrylic dispersions, vinyl acetate dispersions, polybutadien,

Solvents: All kind of solvents like alcohols, esters, ketones, glycolethers, aliphatics, amines and many more

Polymers: Thermoplastic acrylates, styrene acrylic and pure acrylic dispersions, vinyl acetate dispersions, polybutadien

For applications in the field of
Paper/packaging
Bookbinding/graphic design
Wood/furniture industry
Construction industry, including floors, walls and ceilings
Electronics
Automotive industry, aviation industry
Textile industry
Adhesive tapes, labels
Hygiene
Household, recreation and office

Unique Adhesives

Beardow Adams GmbH
Vilbeler Landstraße 20
D-60386 Frankfurt/M.
Phone +49 (0) 69-4 01 04-0
Fax +49 (0) 69-4 01 04-1 15
www.beardowadams.com

Member of IVK

Company

Year of formation
1875

Sales channels
direct and agencies

Contact partners
Sales and Marketing Management:
Janet Pohl

Marketing Assistant:
Suzie Daniel

Range of Products

Types of adhesives
Hot melt adhesives
Reactive adhesives
Dispersion adhesives
Casein, dextrin and starch adhesives
Pressure-sensitive adhesives

For applications in the field of
Paper converting/packaging/labelling
Bookbinding/graphic design
Wood/furniture industry
Electronics
Mechanical engineering and equipment
construction
Automotive industry
Adhesive tapes, labels

Berger-Seidle GmbH

Parkettlacke · Klebstoffe · Bauchemie

Maybachstraße 2
D-67269 Grünstadt/Weinstraße
Phone +49 (0) 63 59-80 05-0
Fax +49 (0) 63 59-80 05-50
Email: info@berger-seidle.de
www.berger-seidle.de

Member of IVK

Company

Year of formation
1926

Size of workforce
110

Ownership structure
100 % subsidiary to Phil. Berger GmbH

Sales channels
Distributors and sales partners/representatives on each country

Contact partners
Management & Sales:
Markus M. Adam

Application technology:
Dr. Wolfgang Kahlen

Further information
www.berger-seidle.de

Range of Products

Types of adhesives
Reactive adhesives
Solvent-based adhesives
Dispersion adhesives
PU-adhesives

Types of sealants
Acrylic sealants
PUR sealants
MS/SMP sealants

For applications in the field of
Wood/furniture industry
Construction industry, including floors,
walls and ceilings

Bilgram Chemie GmbH

Torfweg 4
D-88356 Ostrach
Phone +49 (0) 7585 9312-0
Fax +49 (0) 7585 9312-94
Email: info@bilgram.de
www.bilgram.de

Member of IVK

Company

Year of formation
1971

Size of workforce
250

Ownership structure
family-owned

Sales channels
direct sales, wholesale trade, trade partners

Contact partners
Application technology:
Email: roland.opferkuch@bilgram.de
Sales:
Email: emmanuel.giotis@bilgram.de

Range of Products

Types of adhesives
Reactive adhesives
Dispersion adhesives

Raw materials
Fillers
Solvents

For applications in the field of
Paper/packaging
Bookbinding/graphic design
Wood/furniture industry
Construction industry, including floors, walls and ceilings

Biesterfeld Spezialchemie GmbH

Biesterfeld Spezialchemie GmbH
Ferdinandstraße 41
D-20095 Hamburg
Fax +49 (0) 40 32008-671
Email: m.liebenau@biesterfeld.com
www.biesterfeld.com

Company

Year of formation
1988

Size of workforce
262

Managing partners
Peter Wilkes, Thomas Arnold

Ownership structure
100% subsidiary of Biesterfeld AG

Subsidiaries
in more than 20 countries throughout Europe

Contact partners
Management:
Dr. Martin Liebenau, Business Manager CASE, m.liebenau@biesterfeld.com

Range of Products

Types of adhesives
Hot melt adhesives
Reactive adhesives
Solvent-based adhesives
Dispersion adhesives
Vegetable adhesives, dextrin and starch adhesives
Glutine glue
Pressure-sensitive adhesives

Types of sealants
Acrylic sealants, Butyl sealants
Polysulfide sealants, PUR sealants
Silicone sealants, MS/SMP sealants, Other

Raw materials
Additives: Antioxidants, CMC, Deaerators, Defoamers, Dispersants, Emulsifiers, Epoxy hardeners, Flame retardants, Flow and Leveling Agents, PUR-catalysts, Rheology modifiers, UV-stabilizers, Wetting agents, Xantan Gum
Resins: Acrylic Resins, Polybutens, Polyester-polyols, Polyetherpolyols, Pre-Polymers
Polymers: MS Polymers
Starch: Yellow and white dextrin, Starch esters, Starch ethers

For applications in the field of
Paper/packaging
Bookbinding/graphic design
Wood/furniture industry
Construction industry, including floors, walls and ceilings
Electronics
Textile industry
Adhesive tapes, labels

BOLTON ADHESIVES

Bison International B.V.
Dr. A.F. Philipsstraat 9
NL-4462 EW Goes
Phone +31 (0) 88 3 235 700

Fax + 31 (0) 88 3 235 800
Email: info@boltonadhesives.com
www.boltonadhesives.com
www.bison.net, www.griffon

Headquarter: Bolton Adhesives
Adriaan Volker Huis – 14th floor
Oostmaaslaan 67, NL-3063 AN Rotterdam

Member of IVK

Company

Year of formation
1938

Size of workforce
Bolton Adhesives: > 750

Subsidiaries
Bison International, Zaventem (B)
Griffon France S.a.r.l, Compiègne (F)
Productos Imedio S.A., Madrid (E)

Sales channels
Technical Trade, Hardwarestore, Food
Distribution, Paper-, Office- & Stationery
Trade, Drugstores, Departmentstores

Contact partners
Management:
Rob Uytdewillegen, Danny Witjes

Further information
Technology:
Wiebe van der Kerk, Mariska Grob,
Charlotte Janse, John van Duivendijk

Sales:
Professional Business: Egbert Willemse
DIY: Frank Heus

Range of Products

Types of adhesives
Hot melt adhesives
Reactive adhesives
Solvent-based adhesives
Dispersion adhesives
Pressure-sensitive adhesives

Types of sealants
Acrylic sealants
PUR sealants
Silicone sealants
MS/SMP sealants

For applications in the field of
Paper/packaging
Wood/furniture industry
Construction industry, including floors, walls
and ceilings
Electronics
Automotive industry, aviation industry
Adhesive tapes, labels
Household, recreation and office

BODO MÖLLER CHEMIE
Engineer chemistry

Bodo Möller Chemie GmbH
Senefelderstraße 176
D-63069 Offenbach am Main
Phone +49 (0) 69-83 83 26-0
Fax +49 (0) 69-83 83 26-199
Email: info@bm-chemie.de
www.bm-chemie.com

Company

Year of formation
1974

Employees
Over 200

Managing partners
Korinna Möller-Boxberger, Frank Haug, Jürgen Rietschle

Ownership structure
Family owned

Regions
Germany, Austria, Slovenia, Switzerland, France, Benelux, United Kingdom, Ireland, Denmark, Sweden, Norway, Finland, Estonia, Poland, Lithuania, Latvia, Czech Republic, Slovakia, Hungary, Croatia, Russia, India, China, Southern Africa, Sub Sahara Region, Kenya, Egypt, Morocco, Middle East, Israel, USA and Mexico.

Sales channels
Own sales and logistics structures in Europe, Africa, Asia and America.

Contact partners
Management:
info@bm-chemie.de

Application technology and sales:
info@bm-chemie.de

Further information
Leading supplier for specialty chemicals and partner for industrial high performance adhesives, casting resins and engineering plastics in Europe, Africa, Asia and America with more than 40 years experience in various applications in all fields of the processing industry. Bodo Möller Chemie has its own adhesives application laboratory and is certified for aviation and railway.

Range of Products

Types of adhesives
Epoxy resin adhesives, polyurethane adhesives, methacrylate adhesives, silicone adhesives, hot melt adhesives, reactive adhesives, solvent-based adhesives, dispersion adhesives, pressure-sensitive adhesives, MS polymers, polycondensation adhesives, UV-curing adhesives, spray adhesives, anaerobic adhesives, cyanoacrylates

Types of sealants
Acrylic sealants, butyl sealants, polysulfide sealants, PUR sealants, silicone sealants, MS/SMP sealants

Raw materials
Additives: stabilizers, antioxidants, rheology modifiers, tackifier, softener, thickener, dispersing agent, flame retardant, pigments, light stabilizers: HALS and UV stabilizers, crosslinkers

Fillers: barium sulphate, dolomite, kaolin, calcium carbonate, zinc, talc, aluminium oxide

Resins: acrylate dispersions, polyurethane dispersions, epoxy resins, rosin resins, reactive diluent, cobalt-free drying agent

Polymers: Formulated polymers EP, PU, PA

For applications in the field of
Paper/packaging, bookbinding/graphic design, wood/furniture industry, construction industry, including floors, walls and ceilings, electronics, mechanical engineering and equipment construction, automotive industry, aviation industry, textile industry, adhesive tapes, labels, hygiene, household, recreation and office, adhesive applications for lightweight construction, composite bonding

Bona AB
Murmansgatan 130, Box 210 74
S-20021 Malmö
Phone +46 40 38 55 00, Fax +46 40 18 16 43
Email: bona@bona.com

Adhesive Production
Bona GmbH Deutschland
Jahnstraße 12, D-65549 Limburg
Phone +49 (0) 64 31-4 00 80

Member of IVK

Company

Year of formation
1919

Size of workforce
600

Ownership structure
privat owned stock cooperation

Subsidiaries
Austria: Bona Austria GmbH
Phone +43 662 66 19 43-0
Belgium: Bona NV
Phone +32 2 721 2759
Brazil: BonaKemi Pisos de Madeira
do Brasil Ltda
Phone +55 41 3233 5983
Bona AB Branch Panama
Phone +507 227 2799
Bona Trading (Shanghai) Co., Ltd
Phone +86 10 67 72 8301
Czech Republic & Slovak Republic:
Bona CR spol. s.r.o.
Phone CR +420 236 080 211
Phone SR +421 265 457 161
France: Bona France
Phone: +33 3 88 49 18 60
Germany
Bona Vertriebsges. mbH Deutschland
Phone +49 6431 4008 0
The Netherlands: Bona Benelux BV
Phone +31 23 542 1864
Poland: Bona-Polska Sp. z.o.o.
Phone +48 61 816 34 60/61
Romania: Bona S.r.l.
Phone +40 31 405 75 93

Range of Products

Types of adhesives
Reactive adhesives
Dispersion adhesives

For applications in the field of
Construction industry, for wood floors and
LVT's

Singapore: BonaFar East & Pacific Pte Ltd
Phone +65 6377 1158
Spain/Portugal: Bona Ibérica
Phone +34 916 825522
Italy: Biffignandi spa – Via Circonvallazione
Est 2/6 – Cassolnovo (PV) Italia
Phone 0381 920111
Hungary: M.L.S. Magyarország Kft. -
2310 Szigetszentmiklós, Sellő u. 8.
HUNGARY - Phone/Fax: (06 24) 525 400
United Kingdom: Bona Limited
Phone +44 1908 525 150
United States: Bona US
Phone +1 303 371 1411

Contact partners
Management
Dr. Thomas Brokamp (Production, R & D)

Product Management:
Torben Schuy, Thomas Hallberg

Bostik GmbH
An der Bundesstraße 16
D-33829 Borgholzhausen
Phone +49 (0) 54 25-8 01-0
Email: info.germany@bostik.com
www.bostik.de

Member of IVK, FEICA

Company

Year of formation
1889

Size of workforce
400

Subsidiaries
MEM Bauchemie GmbH, Bostik Austria,
Debratec Schwepnitz

Sales channels
Construction distribution, DIY,
industry

Contact partners
Management:
Olaf Memmen, Managing Director
Richard Ricpe, Business Manager Construction
Norbert Uniatowski, Business Manager FlexLam
Frank Mende, R&D Director
Bernd Köhler, Supply Chain Director
Dr. Michael Nitsche, Production Director

Further information
With annual sales of € 2.0 billion, the company employs 6,000 people and has a presence in more than 50 countries. For the latest information, visit www.bostik.com

Range of Products

Types of adhesives
Hot melt adhesives
Reactive adhesives
Solvent-based adhesives
Dispersion adhesives
Vegetable adhesives, dextrin and starch adhesives
Pressure-sensitive adhesives
Polymer modified binders
SMP adhesives

Types of sealants
Acrylic sealants
Butyl sealants
PUR sealants
Silicone sealants
MS/SMP sealants

Raw materials
Additives, Fillers, Resins, Solvents
Polymers, Starch

For applications in the field of
Paper/packaging
Bookbinding/graphic design
Wood/furniture industry
Construction industry, including floors, walls and ceilings
Mechanical engineering and equipment construction
Automotive industry, aviation industry
Textile industry
Hygiene
Household, recreation and office
Flexible Laminating

Botament
Systembaustoffe
GmbH & Co. KG

Tullnerstraße 23
A-3442 Langenrohr
Phone +43 (0) 22 72-6 74 81
Fax +43 (0) 22 72-6 74 81-35
Email: info@botament.at
www.botament.at

Member of IVK, FCIO

Company

Year of formation
1993

Size of workforce
15

Ownership structure
GmbH & Co. KG

Sales channels
wholesale

Contact partners
Sales manager:
Prok. Ing. Peter Kiermayr

Application technology and sales:
Karl Prickl

CEO:
DI (FH) Markus Weinzierl

Range of Products

Types of adhesives
Reactive adhesives
Dispersion adhesives
Tile adhesives/natural stone adhesives

Types of sealants
Silicone sealants

For applications in the field of
Construction industry, including floors walls
and ceilings

Brenntag GmbH
Messeallee 11
45131 Essen
Phone +49 (0) 201 / 64 96 - 0
Email: construction@brenntag.de
www.brenntag-gmbh.de

Member of IVK

Company

Year of formation
1874

Size of workforce
1,200

Managing partners
Oliver Rechtsprecher,
Roland Saenger,
Mike Dudjan

Nominal capital
154.5 Mio. Euro (Brenntag AG)

Ownership structure
Listed on the stock exchange (Brenntag AG)

Contact partners
Management:
Alain Kavafyan

Application technology and sales:
Michael Hesselmann
Markus Wolff

Range of Products

Raw materials
Additives:
Accelerators, Adhesion Promoters,
Antioxidants, Biocides, Catalysts, Defoamers, Dispersing Agents, Matting Agents,
Plasticizers, Polyether Amines, Rheology
Modifiers, Silanes, Surfactants, Thickeners,
UV-Stabilisers, Molecular Sieves

Resins:
Acrylic Dispersions and Resins, Acrylic
Monomers, Styrene Acrylics
Epoxy Resins incl. Curing Agents, Reactive
Diluents and Modifiers,
Hydrocarbon Resins
PU-Systems: Polyols and Isocyanates
(aromatic and aliphatic),
Prepolymers, PUR-Dispersions
Silikone Resins and Emulsions

Solvents: all kinds of solvents

Polymers: PMMA, Silicones

For applications in the field of
Paper / packaging
Bookbinding / graphic design
Wood / furniture industry
Construction industry, including floors, walls
and ceilings
Electronics
Automotive industry, aviation industry
Textile industry
Adhesive tapes, labels
Composites

)(BÜHNEN

Bühnen GmbH & Co. KG
Hinterm Sielhof 25
D-28277 Bremen
Phone +49 (0) 4 21-51 20-0
Fax +49 (0) 4 21-51 20-2 60
Email: info@buehnen.de
www.buehnen.de

Member of IVK

Company

Year of formation
1922

Size of workforce
97

Ownership structure
Private ownership

Subsidiaries
BÜHNEN Polska Sp. z o. o.
BÜHNEN B. V., NL
BÜHNEN, AT

Contact person
Managing Director & Shareholder:
Bert Gausepohl

Managing Director:
Jan-Hendrik Hunke

International Sales/Marketing
Valentino Di Candido

Sales GER, AT, CH, NL, PL, Intl.
Jan-Hendrik Hunke

Range of Products

Distribution channels
Direct Sales, Distributors

Hot Melt Adhesives
The product range includes a
variety of different hot melt adhesives
for almost every application.
Available bases:
EVA, PO, POR, PA, PSA, PUR, Acrylate.
Available shapes:
slugs, sticks, granules, pillows, blocks,
cartridges, barrels, drums, bags.

Application Technology
Hot melt tank applicator systems with
piston pump or gear pump, PUR- and POR-
hot melt tank systems, PUR- and POR-bulk
unloader, hand guns for spray and bead
application, application heads for bead,
slot, spray and dot application and special
application heads for individual customer
requirements, hand-operated glue applicators, PUR- and POR glue applicators, wide
range of application accessories, customer-
oriented application, solutions.

Applications
Automotive, Packaging, Display Manufac-
turing, Electronic Industry, Filter Industry,
Shoe Industry, Foamplastic and Textile
Industry, Case Industry, Construction
Industry, Florists, Wood-Processing and
Furniture Industry, Labelling Industry.

BYK
Abelstraße 45
D-46483 Wesel
Phone +49 (0) 281-670-0
Fax +49 (0) 281-6 57 35
Email: info@byk.com
www.byk.com

Member of IVK

Company

Year of formation
1962

Size of workforce
Around 2,300 people worldwide

Managing partners
Dr. Stephan Glander (CEO), Alison Avery
(CFO & NAFTA), Gerd Judith (Asia), Dr. Stefan
Mößmer (Paint Additives), Theodore Williams
(Plastics Additives), Holger Heilmann (Industrial
Applications), Frank R. Wagner (Measuring &
Testing Instruments), Donald Poucher (Oil & Gas
Industry), Dr. Horst Sulzbach (CTO)

Ownership structure
BYK is a member of ALTANA, Germany

Subsidiaries
BYK-Chemie (Germany), BYK (Brazil), BYK Additives
(China), BYK (India), BYK Japan (Japan), BYK Korea
(Korea), BYK Netherlands (Netherlands),
BYK Chemie de México (México), BYK Asia Pacific
(Singapore, Taiwan, Thailand and Vietnam),
BYK Additives (United Kingdom), BYK (U.A.E.),
BYK USA (USA), PolyAd Services (USA)
• Warehouses and representations in
  > 100 countries and regions
• Technical Service Labs in Germany, Brazil, China,
  India, Japan, Korea, México, The Netherlands,
  Singapore, U.A.E., United Kingdom and in the USA
• Production Sites in Germany, China, The
  Netherlands, United Kingdom and in the USA

Sales channels
Worldwide – direct (BYK) and indirect
(agents and distributors)

Close to the customer
BYK places great importance on being close to its
customers and remaining in constant dialog with
them. This is one of the reasons why the company is
represented in more than 100 countries and regions
around the globe. In over 30 technical service
laboratories, BYK offers customers and application
engineers support for concrete questions.

Range of Products

Raw materials
Additives: Wetting and dispersing additives, rheological additives (PU thickener, organically modified
clays, synthetic and natural layered silicates),
defoamers and air release agents, additives to improve substrate wetting, surface slip and levelling,
UV-absorbers, wax additives, anti-blocking additives, conductive additives, nano based additives,
adhesion promoters

For applications in the field of
Paper/packaging
Bookbinding/graphic design
Wood/furniture industry
Construction industry including floors,
walls and ceilings
Electronics
Sealants
Automotive industry, aviation industry
Textile industry
Adhesive tapes, labels
Hygiene
Household, recreation and office

Contact partners for your customers
Mr. Tobias Austermann,
Email: Tobias.Austermann@altana.com
Phone: +49 281-670-28128

Further information about your company
BYK is one of the world's leading suppliers in the
field of additives and measuring instruments.The
coatings, inks and plastics industries are among
the main consumers of BYK additives. Yet with oil
production, the manufacture of care products, paper surface finishing, the manufacture of adhesives
and sealants, or construction chemistry, too, BYK
additives improve the product characteristics and
production processes.

Byla GmbH

Industriestraße 12
D-65594 Runkel
Phone +49 (0) 64 82-91 20-0
Fax +49 (0) 64 82-91 20-11
Email: contact@byla.de
www.byla.de

Member of IVK

Company

Year of formation
1975

Nominal capital
90,000 €

Sales channels
Worldwide

Range of Products

Types of adhesives
Reactive adhesives

For applications in the field of
Wood/furniture industry
Electronics
Mechanical engineering and equipment
construction
Automotive industry, aviation industry

certoplast
Technische Klebebänder
GmbH

Müngstener Straße 10
D–42285 Wuppertal
Phone +49 (0) 20 2-2 55 48-0
Fax +49 (0) 20 2-2 55 48-48
Email: verkauf@certoplast.com
www.certoplast.com

Member of IVK
Member of Afera

Company

Year of formation
1991

Size of workforce
90

Managing partners
P. Rambusch
Dr. R. Rambusch

Subsidiaries
certoplast (Suzhou) Co., Ltd., China
certoplast North America Inc., Las Cruses

Contact partners
Management:
P. Rambusch (General manager)
Dr. R. Rambusch (General manager)

Application technology and sales:
Dr. Andreas Hohmann (Sales manager)

Range of Products

Types of adhesives
Hot melt adhesives
Dispersion adhesives
Pressure-sensitive adhesives

Equipment, plant and components
for adhesive curing
adhesive curing and drying

For applications in the field of
Construction industry, including floors,
walls and ceilings
Electronics
Automotive industry, aviation industry
Adhesive tapes, labels

We create chemistry

expect more ✚

Chemetall GmbH
Trakehner Straße 3
D-60487 Frankfurt/M.
Phone +49 (0) 69-71 65-0
Fax +49 (0) 69-71 65-29 36
www.chemetall.com

Member of IVK

Company

Year of formation
1982

Managing partners
Board of Management:
Christophe Cazabeau

Ownership structure
A Global Business Unit of BASF's
Coatings division

Subsidiaries
> 40 worldwide

Sales channels
CM subsidiaries and specific distributors

Contact partners
Management:
Thomas Willems

Application technology and sales:
Ralph Hecktor
Phone +49 (0) 69-71 65-24 46

Further information
See website: www.chemetall.com
Certification to ISO 9001, EN 9100,
ISO 14001

Range of Products

Types of adhesives
Hot melt adhesives
Reactive adhesives

Types of sealants
Polysulfide sealants
PUR sealants
Other epoxy

For applications in the field of
Electronics
Mechanical engineering and equipment
construction
Automotive industry, Aviation industry

Chemische Fabrik Budenheim KG

Rheinstraße 27
D-55257 Budenheim
Phone +49 (0) 6139-89-0
Email: info@budenheim.com
www.budenheim.com

Member of IVK

Company

Year of formation
1908

Size of workforce
1.000

Managing partners
Dr. Harald Schaub
Dr. Stefan Lihl

Ownership structure
Belongs to the Oetker Group, privately owned

Contact partners
Management:
Email: coatings@budenheim.com

Applications technology and sales:
Email: coatings@budenheim.com

Furher information
www.budenheim.com/clip4coatings

Range of Products

Raw materials
Additives
Fillers
Flame Retardants

For applications in the field of
Wood/furniture industry
Construction industry, including floors, walls and ceilings
Automotive industry, aviation industry
Textile industry

**SMART CHEMISTRY
WITH CHARACTER.**

CHT Germany GmbH
Bismarckstraße 102
D-72072 Tübingen
Phone +49 7071 154-0
Fax +49 7071 154-290
Email: info@cht.com
www.cht.com

Member of IVK

Company

Year of formation
1953

Size of workforce
2,200 worldwide

Sales channels
More than 29 CHT affiliates and agencies
worldwide

Management
Dr. Frank Naumann (CEO)
Dr. Bernhard Hettich (COO)
Axel Breitling (CFO)

Contact partners
Application technology and sales:
Dennis Seitzer (Textile, F&E Polymers)

Range of Products

Types of adhesives
Powder adhesives
Reactive adhesives
Solvent-based adhesives
Dispersion adhesives
High solid adhesives
Pressure-sensitive adhesives
Acrylic adhesives
PUR adhesives
Silicone adhesives
Thermo-activated adhesives

Types of sealants
Silicone sealants

Raw materials
RTV-1 / RTV-2
LSR silicones
Acrylic dispersions
PU dispersions

Additives:
Adhesion promoters and primers
Rheology additives and thickeners
Release agents
Crosslinker, chain extender

For applications in the field of
Construction industry
Electronics
Mechanical engineering and equipment
Automotive industry, aviation industry
Textile industry / Technical textiles
Paper and packaging
Pressure sensitive tapes and labels
Flock
Mould making

CnP Polymer GmbH

Schulteßdamm 58
D-22391 Hamburg
Phone +49 (0) 40-53 69 55 01
Fax +49 (0) 40-53 69 55 03
Email: info@cnppolymer.de
www.cnppolymer.de

Member of IVK

Company

Year of formation
1999

Ownership structure
private

Sales channels
own sales force

Contact
Christoph Niemeyer

Range of Products

Raw materials
Polymers:
SIS, SBS, SEBS, SSBR
PIB

Resins:
Hydrocarbon resins

For applications in the field of
Paper/packaging
Bookbinding/graphic design
Wood/furniture industry
Construction industry, including floors,
walls and ceilings
Automotive industry, aviation industry
Textile industry
Adhesive tapes, Labels
Hygiene
Household, recreation and office

Coim Deutschland GmbH
Novacote Flexpack Division
Schnackenburgallee 62
D-22525 Hamburg
Phone +49 (0) 40-85 31 03-0
Fax +49 (0) 40-85 31 03-69
Email: info@coimgroup.com
www.coimgroup.com

Member of IVK

Company

Year of formation
as COIM Group in 1962

Ownership structure
private owned company

Subsidiaries
COIM operates through a network of production sites, commercial companies and agencies located all over the world

Sales channels
The Novacote Division is part of the COIM Group, dedicated to developing and supplying adhesives and coatings, mainly for the Flexible Packaging market.

Contact partners
Management:
Frank Rheinisch

Application technology:
Oswald Watterott

Sales:
Joerg Kiewitt

Further information
The Novacote Division is part of the Coim Group, dedicated to developing and supplying adhesives and coatings, mainly for the Flexible Packaging market. During recent years the Novacote Division grew rapidly both in terms of Business and Organization. With respect to the global organization the Novacote Technology Center is located in Hamburg, Germany as R&D Centre for Packaging.

Range of Products

Types of adhesives
Reactive adhesives
Solvent-based adhesives
Dispersion adhesives

Types of sealants
Acrylic sealants
Other

Raw materials
Resins
Polymers

For applications in the field of
Paper/packaging
Adhesive tapes, labels
Hygiene
Insulation materials
Solar panels

Collall B.V.

Electronicaweg 6
NL – 9503 EX Stadskanaal
Phone +31 (0) 599-65 21 90
Fax +31 (0) 599-65 21 91
Email: info@collall.nl
www.collall.nl

Member of VLK

Company

Year of formation
1949

Size of workforce
25

Ownership structure
family-owned

Contact partners
Management:
Patrick van Rhijn

Range of Products

Types of adhesives
Solvent-based adhesives
Dispersion Adhesives
Vegetable adhesives, dextrin and
starch adhesives

For applications in the field of
Household, hobby and office
Bookbinding/graphis design
Wood/furniture industry

Additional
supplier of various creative materials
for school and hobby

Collano AG
CH-6203 Sempach Station
Phone +41 41 469 92 75
Email: info@collano.com
www.collano.com

Member of FKS

Company

Year of formation
1947

Size of workforce
20

Sales channels
Direct sales or distributors

Contact partners
Management:
Mike Gabriel

Sales:
Mike Gabriel
Phone: +41 41 469 92 75

Range of Products

Types of adhesives
Hot melt adhesives
Reactive adhesives
Dispersion adhesives

For applications in the field of
Wood/furniture industry
Construction industry

Coroplast
Fritz Müller
GmbH & Co. KG

Wittener Straße 271
D-42279 Wuppertal
Phone +49 (0) 2 02-26 81-0
Fax +49 (0) 2 02-26 81-3 80
Email: coroplast@coroplast.de
www.coroplast.de

Member of IVK

Company

Year of formation
1928

Size of workforce
5,700

Sales channels
Wholesale and industry

Contact partners
Management:
N. Mekelburger
M. Söhngen
W. Berns
T. Kämmerer

Further information

Coroplast acts in 3 business units:
• tapes
• cables & wires
• cable assemblies

Range of Products

For applications in the field of
Paper/packaging
Wood/furniture industry
Construction industry, including floors,
walls and ceilings
Electronics
Mechanical engineering and equipment
construction
Automotive industry, aviation industry
Household, recreation and office

cph Deutschland Chemie GmbH

Heinz-Bäcker-Straße 33
D-45356 Essen
Phone +49 (0) 2 01- 81 40 60
Email: service@cph-group.com
www.cph-group.com

Member of IVK

Company

Year of formation
1975

Size of workforce
ca. 210 (cph group)

Managing partners
CEO Dr. Gerwin Schüttpelz

Subsidiaries
Russia, Portugal, South Africa

Range of Products

Types of adhesives
Hot melt adhesives
Dispersion adhesives
Vegetable adhesives, dextrin and starch adhesives, casein
Pressure-sensitive adhesives

For applications in the field of
Labelling
Paper/packaging
Wood/furniture industry
Adhesive tapes, labels
Hygiene

Covestro AG

D-51373 Leverkusen
Phone +49 (0) 214-6009 3080
Email: adhesives@covestro.com
www.adhesives.covestro.com

Member of IVK

Company

Year of formation
2015

Size of workforce
16,800 (31. December 2018)

Contact partners
Marketing Europe/Business Development
Phone +49 (0) 214-6009 3080
Email: adhesives@covestro.com
www.adhesives.covestro.com

Range of Products

Raw materials
Polyurethane-Dispersions (Dispercoll® U)
Hydroxylpolyurethanes (Desmocoll®, Desmomelt®)
Polyisocyanates (Desmodur®)
Isocyanate-Prepolymers (Desmodur®, Desmoseal®)
Silanterminated Polyurethanes (Desmoseal® S)
Polyesterpolyols (Baycoll®)
Polyetherpolyols (Desmophen®, Acclaim®)
Polychloroprene-Dispersions (Dispercoll® C)
Halogenated Polyisoprenes (Pergut®)
Silicon dioxide-nanoparticle dispersions (Dispercoll® S)

CTA GmbH
Voithstraße 1
D-71640 Ludwigsburg
Phone +49 (0) 71 41 - 29 99 16 - 0
Email: info@cta-gmbh.de
www.cta-gmbh.de

Member of IVK

Company

Year of formation
2005

Size of workforce
200

Managing partners
Martin Kummer

Ownership structure
Member of Tubex Holding GmbH

Sales channels
direct

Contact partners
Management:
Martin Kummer

Application technology and sales:
Sales: Franco Menchetti
R + D: Dr. Dirk Buchholz

Further information
The core business of CTA GmbH is a wide
range of contract services in the fields of:
- Product manufacturing which includes the
  developing or improvement of composites
  and the mixing of products according to
  customers' recipes and parameters
- The filling of almost all viscous fluid chemical
  products into various kinds of primary and
  secondary packaging – like all kind of tubes,
  cartridges, bottles, pouches, cartons, blister
  packs and others – which are released
  and approved to the best suitability of the
  product and its adequate packaging.

Range of Products

Types of adhesives
Solvent-based adhesives
Dispersion adhesives
Glutine glue
Pressure-sensitive adhesives

Raw materials
Additives, Fillers, Resins, Solvents

Types of sealants
Silicone sealants
MS / SMP sealants
Other

Equipment, plant and components
measuring and testing

For applications in the field of
Paper / packaging
Wood / furniture industry
Construction industry, including floors, walls
and ceilings
Electronics
Wind energy
Mechanical engineering and equipment
construction
Automotive industry, aviation industry
Household, recreation and office

- Developing the appropriate packaging in
  accordance to the needs of the product and
  its marketing aspects.
- The packaging assembling to the point of
  sale, supply and distribution logistics offers
  a full contract service package to different
  branches of industries and businesses.

Cyberbond Europe GmbH – A H.B. Fuller Company
Werner-von-Siemens-Straße 2, D-31515 Wunstorf
Phone +49 (0) 50 31-95 66-0, Fax +49 (0) 50 31-95 66-26
Email: info@cyberbond.de, www.cyberbond.eu
Member of IVK

Company

Year of formation
1999

Size of workforce
24

Managing partners
Ulrich Lipper, Holger Bleich, James East, Robert Martsching

Nominal capital
50,000 EUR

Ownership structure
H.B. Fuller

Subsidiaries
Cyberbond France SARL, France
Cyberbond Iberia, Spain
Cyberbond CS s. r. o., Czech Republic

Sales channels
Direct to the industry and via exclusive nationwide distributors as well as special Private Label accounts

Contact partners
Marketing and Sales:
Ulrich Lipper

Application technology:
Dr. Lars Hoyer

Further information
Cyberbond –
The Power of Adhesive Information
IATF 16949
ISO 13485
ISO 9001
ISO 14001

Range of Products

Offered adhesives
Cyanoacrylates
Anaerobic Adhesives and Sealants
UV and Light Curing Adhesives
Additional programme consisting of:
Primers, Activators, D-Bonders and Dosing Aids

Dosing equipment
LINOP Modular Dosing System for
1K Reactive Adhesives
LINOP UV LED Curing System

Products are used in
Automotive and automotive sub supplier industry
Electronic industry
Aviation industry
Elastomer/plastic/metal working industry
Machine tool industry
Medical industry
Shoe industry
DIY, Hobby and office
Wood/furniture industry

DEKA
Kleben & Dichten
GmbH (Dekalin®)

Gartenstraße 4
D-63691 Ranstadt
Phone +49 (0) 60 41-82 03 80
Fax +49 (0) 60 41-82 12 ??
Email: info@dekalin.de
www.dekalin.de

Member of IVK

Company

Year of formation
1907 DEKALIN
1999 DEKA

Size of workforce
together > 125

Ownership structure
family owned

Sales channels
industry, manufacturing,
technical wholesalers

Contact partners
Management:
Michael Windecker

Range of Products

Types of adhesives
Solvent-based adhesives
Dispersion adhesives

Types of sealants
Butyl sealants
PUR sealants
MS/SMP sealants

For applications in the field of
Wood/furniture industry
Construction industry, including floors,
walls and ceilings
Mechanical engineering and equipment
construction
Automotive industry, aviation industry
Household, recreation and office
Caravan, camper and mobile home,
HVAC, Air duct sealing

Delo Industrial Adhesives

DELO-Allee 1
D-86949 Windach
Phone +49 (0) 81 93-99 00-0
Fax +49 (0) 81 93-99 00-1 44
Email: info@DELO.de
www.DELO-adhesives.com

Member of IVK

Company

Year of formation
1961

Size of workforce
780

Managing board
Dr. Wolf-Dietrich Herold
Sabine Herold
Robert Saller

Subsidiaries
Global distributors, subsidiaries in the USA, China, Singapore and Japan and representative offices in Taiwan, Malaysia and South Korea

Sales channels
Traders and direct

Contact partners
Application technology and sales:
Robert Saller, Managing director

Further information
DELO is a leading manufacturer of industrial adhesives with its headquarters in Windach near Munich, Germany. In the fiscal year 2019, the company generated sales revenues of EUR 156 million. The company supplies customized adhesives and associated technology for high-tech industries such as automotive, consumer electronics and optoelectronics. DELO's customers include Bosch, Daimler, Huawei, Osram, Siemens and Sony.

Range of Products

Types of adhesives
Dual-curing adhesives
Light-curing and light-activated acrylates and epoxies
One- and two-component epoxy resins
Electrically conductive adhesives
Methacrylates
Polyurethanes
Cyanoacrylates

Types of sealants
Acrylic sealants
PUR sealants
Silicone sealants
Other

Equipment, plant and components
Spot and area curing lamps
Jet-valves for micro-dispensing
Flexible foil cartridges for bubble-free dispensing

For applications in the field of
Automotive
Consumer and industrial electronics
Optics and optoelectronics
Mechanical engineering
Aviation
White goods

Distona AG

Hauptplatz 5
CH-8640 Rapperswil
Phone +41 (0) 55 533 00 50
Fax +41 (0) 55 533 00 51
Email: info@distona.ch
www.www.distona.ch

Member of FKS

Company

Year of formation
2014

Size of workforce
5 employees

Managing partners
Daniel Altorfer
David Nipkow

Range of Products

Raw materials
Additives:
Stabilizers, biocides, flame retardants,
silane, siloxane, rheology modifiers

Fillers:
Porous Glas Powder, Bentonite, Kaolinite,
Chalk, Talcum, Zeolite

Resins:
Epoxide Technologies, Isocyanates, Polyols,
Acrylic monomers

Solvents Bio-solvents, conventional solvents

DKSH GmbH
Baumwall 3
20459 Hamburg
Phone +49 (0) 40-37 47-340
Fax +49 (0) 40-37 47-380
Email: info.ham@dksh.com
www.dksh.de

Member of IVK

Company

Year of formation
1975

Size of workforce
50

Managing Director
Thomas Sul

Ownership structure
DKSH Group

Subsidiaries of DKSH Group
33,000 employees/
net sales CHF 11,3 billion

Sales channels
Own sales force

Sales Manager
Sven Thomas

Further information
DKSH is the leading Market Expansion Services provider with a focus on Asia. The Group helps other companies and brands to grow in the Consumer Goods, Healthcare, Performance Materials and Technology sectors. DKSH's portfolio of services includes sourcing, market insights, marketing and sales, distribution and logistics as well as after-sales services. Publicly listed on the SIX Swiss Exchange, the Group operates in 35 markets with 33,000 specialists, generating net sales of CHF 11.3 billion in 2018. The DKSH Business Unit Performance Materials distributes specialty chemicals and ingredients for

Range of Products

Raw materials
Wide range of specialties for Epoxi and PU
Aliphatic isocyanates
High-molecular weight co-polyesters
Liquid isoprene rubbers
Moisture scavengers (PTSI)
Oxetanic reactive diluents and plasticizers
Resins: Co- Polyester, Vinyl, Polyamide-Imide, Acrylic, Ketone, Oxetanic Resins
Heat seal lacquers
Adhesion promoters and primers Polyolefin, Polyolefindispersions, Polyester Lacquers
Plasticizers
Electrical and thermal conductive additives

For applications in the field of
Paper/Pharma/Food packaging Bookbinding/graphic design Wood/furniture industry
Construction industry, including floors, walls and ceilings
Electronics
Automotive industry, aviation industry
Adhesive tapes, labels
Hygiene

food, pharmaceutical, personal care and various industrial applications. With 29 innovation centers and regulatory support worldwide, we create cutting edge formulations that comply with local regulations.

Drei Bond GmbH
Carl-Zeiss-Ring 13
85737 Ismaning, Germany
Phone +49 (0) 89-962427 0
Fax +49 (0) 89-962427 19
Email: info@dreibond.de
www.dreibond.de

Member of IVK

Company

Year of formation
1979

Number of employees
52

Partners
Drei Bond Holding GmbH

Share capital
€ 50,618

Subsidiaries
Drei Bond Polska sp. z o.o. in Kraków

Distribution channels
Directly to the automotive industry
(OEM + tier 1 / tier 2); indirectly
via trading partners as well as select private
label business

Contacts
Management:
Mr. Thomas Brandl

Application engineering, adhesive and
sealants:
Johanna Wiethaler, Christian Eicke

Application engineering, metering technology:
Sebastian Schmidt, Marko Hein

Adhesive and sealant sales:
Thomas Hellstern, Harald Jost, Christian Eicke

Metering technology sales:
Sebastian Schmidt, Marko Hein

Additional information
Drei Bond is certified according to ISO 9001-
2015 and ISO 14001-2015

Range of Products

Types of adhesives/sealants
- Cyanoacrylate adhesives
- Anaerobic adhesives and sealants
- UV-light curing adhesives
- 1C/2C epoxy adhesives
- 2C MMA adhesives
- 1C MS hybrid adhesives and sealants
- 1C synthetic adhesives and sealants
- 1C silicone sealants

Complementary products:
- Activators, primers, cleaners

Equipment, systems and components
- Drei Bond Compact metering systems →
  semi-automatic application of adhesives and
  sealants, greases and oils
  Metering technology: pressure/time and
  volumetric
- Drei Bond Inline metering systems →fully
  automated application of adhesives and
  sealants, greases and oils
  Metering technology: pressure/time and
  volumetric
- Drei Bond metering components:
  Container systems: tanks, cartridges, drum
  pumps
  Metering valves: progressive cavity pumps,
  diaphragm valves, pinch valves, spray
  valves, rotor spray

For applications in the following fields
- Automotive industry/automotive suppliers
- Electronics industry
- Elastomer/plastics/metal processing
- Mechanical and apparatus engineering
- Engine and gear manufacturing
- Enclosure manufacturing (metal and plastic)

DuPont Transportation & Industrial

Wolleraustrasse 15a
CH-8807 Freienbach
Switzerland
www.dupont.com

Member of IVK

Company

About DuPont
DuPont (NYSE: DD) is a global innovation leader with technology-based materials, ingredients and solutions that help transform industries and everyday life. Our employees apply diverse science and expertise to help customers advance their best ideas and deliver essential innovations in key markets including electronics, transportation, building and construction, health and wellness, food and worker safety. More information can be found at www.dupont.com

Contact
Dr. Andreas Lutz
R&D Director Adhesives
DuPont Transportation & Industrial
Email: andreas.lutz@dupont.com
Phone: +41 44 728 35 59
Mobile: +41 79 347 67 69

Range of Products

About DuPont
Transportation & Industrial
DuPont Transportation & Industrial (T&I) delivers a broad range of technology-based products and solutions to the transportation, electronics, healthcare, industrial and consumer markets. T&I partners with customers to drive innovation by utilizing its expertise and knowledge in polymer and materials science. T&I works with customers throughout the value chain to enable material systems solutions for demanding applications and environments. For additional information about DuPont Transportation & Industrial, visit the DuPont businesses page.

Dymax Europe GmbH
Kasteler Straße 45
D-65203 Wiesbaden
Phone +49 (0) 611 962 7900
Fax +49 (0) 611-962 9440
Email: info_DE@dymax.com
www.dymax.com

Member of IVK

Company

Year of formation
1995

Size of workforce
250 + worldwide

Ownership structure
Dymax Corporation, USA

Sales channels
Direct and Distributors

Contact partners
Managing Director:
Christoph Gehse

Technical Manager:
Wolfgang Lorscheider

Further information
At Dymax we combine our product offering
of oligomers, adhesives, coatings, dispensing
systems, and curing equipment with our
expert knowledge of light-cure technology.

Range of Products

Types of adhesives
UV and light-curable adhesives
Temporary masking resins
Conformal coatings
Potting materials
Encapsulants
FIP/CIP- gaskets
In addition: Materials with secondary
moisture- and heat cure options, adhesive
activators.

For applications in the field of
Medical
Orthopedic implants
Electronics
Automotive
Aerospace
Optics
Glass
Optical Bonding

Additional Products
UV-Spot and flood lamps (Broadband and LED)
UV-Conveyors
Radiometers
Dispensing equipment
Technical consulting

ADHESIVES • COATINGS

Eluid Adhesive GmbH
Heinrich-Hertz-Straße 10
D-27283 Verden
Phone +49 (0) 42 31-3 03 40-0
Fax +49 (0) 42 31-3 03 40-17
Email: info@eluid.de
www.eluid.de

Member of IVK

Company

Range of Products

Year of formation
1932

Size of workforce
7

Managing partners
Andreas May

Ownership structure
100 % Private

Contact partners
Andreas May
Karin Münker

Sales channels
Europe: through our own sales force,
traders and agents
Worldwide: traders and agencies

Types of adhesives
Acrylate, Styrene-Acrylate, Polyurethane-
and Vinyl acetate dispersions
(A, RA, SA, PUD, PVAC, VAE) for adhesives,
varnish and coatings
PSA Dispersions
PVOH adhesives
Dextrin-, Casein- and starch adhesives
APAO, EVA, PSA, PO, PUR Hotmelts

For applications in the field of
Coatings for films (printable)
Coatings for films (Soft Touch)
Bookbinding/graphic arts
Paper and Converting
Folding boxes
Packaging
Envelope industry
Protection of books
Adhesives for insulation technology
Tapes (single and double sided) / Labels
Flexible packaging
High-gloss film laminating
Pressure sensitive adhesives for film
(removable/permanent)
Heat Seal Products
Wood adhesives D2 / D3
Foam processing industry
Protection foil
Safety documents
Nonwoven industry
Wallpaper industry
Textile industry
Transformerboards

emerell
the future of your production

Emerell AG
Neulandstrasse 3
6203 Sempach Station
Switzerland
Phone +41 41 469 91 00
Email: info@emerell.com
www.emerell.com

Member of FKS

Company

Emerell is the first independent manufacturing partner for industrial manufacturers, processors and distributors of polymer specialties and high-quality adhesives. As a full custom manufacturer, the company offers a wide range of services to its customers and accompanies them from the first test onto the market launch and further development of their products.

Owner
Adrian Leumann

Contact
Dr. Michael Lang,
Senior Vice President & Head of Business Unit Technical Chemistry;
Phone +41 41 469 93 14;
E-Mail: michael.lang@emerell.com

Subsidiary companies
Emerell GmbH, Buxtehude, Germany

Range of Products

Technologies
Blown film extrusion
Cast film extrusion
Extrusion coating
Polymerisation
Chemical reactions
Mixing techniques for liquid, paste-like and Reactive products

Services
Process engineering
Manufacture and processing
Filling, packing, labelling
Purchase of raw materials and logistics
Storing
Safety management

For applications in the following sectors
Car industry
Aviation industry
Shipping industry
Railway industry
Construction and building industry
Fastening technology
Electronics
Medicine and hygiene
Labels, adhesive tapes, packaging
Textiles
System providers

EMS-CHEMIE AG
Business Unit EMS-GRILTECH
Via Innovativa 1
7013 Domat/Ems
Phone +41 81 632 72 02
Fax +41 81 632 74 02
Email: info@emsgriltech.com
www.emsgriltech.com

Member of FKS

Company

Year of formation
1936 founded as Holzverzuckerungs AG (HOVAG), 1960 renamed into EMSER WERKE AG and finally into EMS-CHEMIE AG in 1981.

Size of workforce
2,939 worldwide in December 2018

Sales channels
Direct Sales Channel, Distributors/Traders

Contact partners
Application technology and sales:
Phone: +41 81 632 72 02, Fax: +41 81 632 74 02
Email: info@emsgriltech.com, www.emsgriltech.com

Contact partners
The business unit EMS-GRILTECH is part of EMS-CHE-MIE AG which belongs the EMS-CHEMIE HOLDING AG. We manufacture and sell Grilon, Nexylon and Nexylene fibers, Griltex hotmelt adhesives, Grilbond adhesion promoters, Primid crosslinkers for powder-coatings and Grilonit reactive diluents. We have developed these materials and additives into excellent specialty products for technically demanding applications. In this way we create added value for our customers supporting their continual improvements.

Thermoplastic hotmelt adhesives
Thermoplastic adhesive products for technical and textile bonding applications are sold under the trade name „Griltex®". EMS-GRILTECH has many years of experience in the manufacture of tailor-made copolyamides and copolyesters for different application fields. The melt temperatures and melt viscosity can be modified over a wide range depending on the different requirements of each application. The adhesives are available as powder in a wide range of grain sizes or as granules. Manufacturing is carried out in our own polymerisation and grinding plants.

Griltex® ES - Bonding of Even Surfaces
Hotmelt adhesives for bonding of metal, plastics, glass and other smooth surfaces are produced under the trade name Griltex® ES.

Range of Products

Types of adhesives
Hot melt adhesives

Types of sealants
Other

Raw materials
Additives
Resins
Polymers

Equipment, plant and components
for conveying, mixing, metering and for adhesive application
measuring and testing

For applications in the field of
Paper/packaging
Wood/furniture industry
Construction industry, including floors, walls and ceilings
Electronics
Mechanical engineering and equipment construction
Automotive industry, aviation industry
Textile industry
Adhesive tapes, labels
Hygiene
Household, recreation and office

EMS-GRILTECH's corporate offices with research laboratory, technical service centre and production facility are located at Domat/Ems, Switzerland. We also have state-of-the-art production locations, application development centers and customer service laboratories in USA, China, Taiwan and Japan.

EMS-GRILTECH is present worldwide either with its own sales companies or represented by agents.

EUKALIN Spezial-Klebstoff Fabrik GmbH
Ernst-Abbe-Straße 10
D-52249 Eschweiler
Phone +49 (0) 24 03-64 50-0
Fax +49 (0) 24 03 64 50 26
Email: eukalin@eukalin.de
www.eukalin.com

Member of IVK

Company

Year of formation
1904

Size of workforce
60 employees

Managing partners
Timm Koepchen
Jan Schulz-Wachler

Ownership structure
family owned

Subsidiaries
EUKALIN Corp. USA

Sales channels
Directly and through dealers

Contact partners
Application technology and sales:
Timm Koepchen

Range of Products

Types of adhesives
Hot melt adhesives
Dispersion adhesives
Dextrin and starch adhesives
Glutine glue
Pressure-sensitive adhesives
Latex adhesives
Cold seal adhesives
Polyurethane Adhesives

Equipment, plant and components
for conveying, mixing, and metering
for adhesive application

For applications in the field of
Paper/packaging
Bookbinding/graphic design
Textile industry
Flexible packaging
Building industry
Tapes + Labels
Labelling

Evonik Industries AG

D-45764 Marl, www.evonik.com/crosslinkers,
www.evonik.com/adhesives-sealants,
www.evonik.com/designed-polymers
D-45764 Marl, www.vestamelt.de
D-64293 Darmstadt, www.visiomer.com
D-45127 Essen, www.evonik.com/polymer-dispersions
www.evonik.com/hanse, www.evonik.com/tegopac
D-63457 Hanau, www.aerosil.com,
www.dynasylan.com, www.evonik.com/fp

Member of IVK

Company

Year of formation
2007

Contact partners
Application technology and sales:

Resource Efficiency
Phone +49 (0) 76 23-91-83 92
(application techn.)
Phone +49 (0) 61 81 59-1 34 76 (sales)
Email: aerosil@evonik.com, fillers.pigments@evonik.com
Phone +49 (0) 23 65-49-48 43
Fax +49 (0) 23 65-49-50 30
Email: adhesives@evonik.com
Phone +49 (0) 23 65-49-43 56 (VESTAMELT®)
        +49 (0) 61 51-18-10 02 (VISIOMER®)
Email: vestamelt@evonik.com, visiomer@evonik.com

Nutrition & Care
Phone +49 (0) 2 01-1 73 21 33 (Sales)
Email: info@polymerdispersion.com,
hanse@evonik.com,
TechService-Tegopac@evonik.com

Company information
Evonik is one of the world leaders in specialty chemicals. The focus on more specialty businesses, customer-oriented innovative prowess and a trustful and performance-oriented corporate culture form the heart of Evonik's corporate strategy. They are the lever for profitable growth and a sustained increase in the value of the company. Evonik benefits specifically from its customer proximity and leading market positions. Evonik is active in over 100 countries around the world. In fiscal 2018, the enterprise with more than 32,000 employees generated sales of € 13.3 billion and an operating profit (adjusted EBITDA) of € 2.15 billion from continuing operations.

Range of Products

Types of adhesives
Hot melt adhesives (VESTAMELT®) (DYNACOLL® S)

Types of sealants
Acrylic sealants (DEGALAN®)

Raw materials
Additives: waxes (VESTOWAX®, SARAWAX®), defoamer (TEGO® Antifoam), wetting agents (TEGOPREN®), thickener (TEGO® Rheo), silica nanoparticles (Nanopox®), silicone rubber particles (Albidur®), methacrylate monomers (VISIOMER®), pyrogenic silicas and metal oxides (AEROSIL®, AEROXIDE®), specialty precipitated silicas (SIPERNAT®), functional silanes (Dynasylan®)

Crosslinkers: speciality resins, aliphatic diamines (VESTAMIN®), aliphatic isocyanates (VESTANAT®)

Polymers: amorphous poly-alpha-olefines (VESTOPLAST®), copolyesters (DYNACOLL®), liquid polybutadienes (POLYVEST®), polyacrylates (DEGALAN®, DYNACOLL® AC), silane-modified polymers (Polymer ST, TEGOPAC®), condensation curing silicones (Polymer OH)

Epoxy Curing Agents: Polycarbamide curtives for HDI trimer (Amicure®), Polyamide and amidoamine curing agents (Ancamide®), Aliphatic & cycloaliphatic amine curing agents (Ancamine®)

For applications in the field of
Paper/packaging
Bookbinding/graphic design
Wood/furniture industry
Construction industry, including floors,
Walls and ceilings
Electronics
Automotive industry, aviation industry
Textile industry
Adhesive tapes, labels
Hygiene
Manufacturing of hotmelt adhesives
Wind energy

Fenos AG

Steinheimer Straße 3
D-71691 Freiberg a.N.
Phone +49 (0) 7141 992249-0
Fax +49 (0) 7141 992249-99
Email: info@fenos.de
www.fenos.de

Member of IVK

Company

Year of formation
2015

Size of workforce
10

Sales channels
Technical Sales and representatives
international

Contact partners
Management:
Dr. Rüdiger Nowack, Dr. Natalia Fedicheva,
Yvonne Steinbach

Application technology and sales:
Dr. Rüdiger Nowack

Further information
Fenos AG is an independant system house
for industrial adhesives and PUR formulations.
All products are developed towards
customers' requirements and processes.

Range of Products

Types of adhesives
Hot melt adhesives
Reactive adhesives
Dispersion adhesives
Pressure-sensitive adhesives

For applications in the field of
Paper/packaging
Wood/furniture industry
Construction industry, including floors, walls
and ceilings
Automotive industry, aviation industry
Textile industry
Adhesive tapes, labels
Hygiene
Household, recreation and office

Fermit GmbH

Zur Heide 4
D-53560 Vettelschoß
Phone +49 (0) 26 45-22 07
Fax +49 (0) 26 45-31 13
Email: info@fermit.de
www.fermit.com

Member of IVK

Company

Year of formation
2008

Size of workforce
15

Ownership structure
100 % subsidiary of Barthélémy/France

Sales channels
sanitary and heating equipm. trade,
industry, whole trade

Contact partners
Management:
Alois Hauk

Further information
former Nissen & Volk, Hamburg

Range of Products

Types of adhesives
Solvent-based adhesives
Resin-based adhesives

Types of sealants
Silicone sealants
MS/SMP sealants
Sealing pastes
Other

For applications in the field of
Sanitary and heating industry
Construction Industry
Mechanical engineering and equipment
construction
Automotive industry, aviation industry

fischer Deutschland Vertriebs GmbH

Klaus-Fischer-Straße 1
D-72178 Waldachtal
Phone +49 (0) 74 43 12-0
www.fischer.de

Member of IVK

Company

Year of formation
1948

Size of workforce
5,200

Ownership structure
Privately owned

Subsidiaries
47 (AE, AR, AT, BE, BR, BG, CN, CZ, DK, DE,
ES, FI, FR, GR, HR, HU, IN, IT, JP, KR, MX,
NL, NO, PH, PL, PT, RO, RU, SE, SG, SK, TH,
TR, US, UK)

Sales channels
DIY, specialized trade

Contact partners
Michael Geiszbühl
Email: Michael.Geiszbuehl@fischer.de

Further information
World market leader in chemical fastening
systems

Range of Products

Types of adhesives
Reactive adhesives
Dispersion adhesives
MS / SMP adhesives
UV curing adhesive
Solvent based adhesives

Types of sealants
Acrylic sealants
Silicone sealants
MS/SMP sealants

For applications in the field of
Wood/furniture industry
Construction industry, including floors,
walls and ceilings
DIY

Follmann GmbH & Co. KG

Heinrich-Follmann-Straße 1
D-32423 Minden
Phone +49 (0) 5 71-93 39-0
Fax +49 (0) 5 71-93 39- 300
Email: sales@follmann.com
www.follmann.com

Member of IVK

Company

Year of formation
1977

Management
Dr. Jörn Küster
Dr. Jörg Seubert

Subsidiaries
OOO Follmann
(Moscow/Russian Federation)
Follmann (Shanghai) Trading Co., Ltd.
(Shanghai/China)
Follmann Chemia Polska (Poznan/Poland)
Sealock Sp. z o.o. (Warsaw/Poland)
Sealock Ltd. (Andover/Great Britain)
Intermelt ZAO (St. Petersburg/Russian
Federation)

Contact partners
Holger Nietschke
Sebastian Drewes
Martin Haupt

Range of Products

Types of adhesives
Hotmelt adhesives
Dispersion adhesives
Reactive hotmelt adhesives
Pressure-sensitive adhesives
Starch and Casein adhesives

For applications in the field of
Paper/packaging
Bookbinding/graphic design
Wood + Furniture industry
Automotive industry
Assembly
Textile industry
Adhesive tapes
Labeling
Upholstery, mattresses
Filtration industry
Abrasive industry

Forbo Eurocol Deutschland GmbH

August-Röbling-Straße 2
D-99091 Erfurt
Phone +49 (0) 3 61-7 30 41-0
Fax +49 (0) 3 61-7 30 41-91
Email: info.erfurt@forbo.com
www.forbo-eurocol.de

Member of IVK, FCIO, VLK

Company

Year of formation
1920

Size of workforce
89

Managing partners
Forbo Beteiligungen GmbH,
D-79539 Lörrach

Nominal capital
2.050.000 EUR

Ownership structure
100 % share holder

Sales channels
direct sales and/or through agents

Contact partners
Management:
Dr. Stefan Vollmuth (Managing Director),
Jochen Schwemmle (Managing Director)

Range of Products

Types of adhesives
Reactive adhesives
Solvent-based adhesives
Dispersion adhesives
Pressure-sensitive adhesives

For applications in the field of
Construction industry, including floors,
walls and ceilings

H.B. Fuller

Connecting what matters.™

H.B. Fuller Europe GmbH
Talacker 50
CH-8001 Zürich
www.hbfuller.com

Member of IVK, FKS, VLK

Company

Global reach
- H.B. Fuller operates three regional headquarters across the globe:
  Americas – St. Paul, Minn., U.S.
  EIMEA – Zurich, Switzerland
  Asia Pacific – Shanghai, China
- Direct presence in 35 countries and customers in more than 100 geographic markets.

European commitment
H.B. Fuller has a network of specialised production sites across Europe serving customers in electronics, disposable hygiene, medical, transportation, aerospace, clean energy, packaging, construction, woodworking, general industries and other consumer businesses.

About H.B. Fuller
Since 1887, H.B. Fuller has been a leading global adhesives provider focusing on perfecting adhesives, sealants and other specialty chemical products to improve products and lives. With fiscal 2018 net revenue of over $3 billion, H.B. Fuller's commitment to innovation brings together people, products and processes that answer and solve some of the world's biggest challenges. Our reliable, responsive service creates lasting, rewarding connections with customers. And, our promise to our people connects them with opportunities to innovate and thrive. For more information, visit us at www.hbfuller.com.

Range of Products

Our Technologies
- Hot Melt
- Polymer and Specialty Technologies
- Reactive Chemistries: Urethane Epoxy Solventless
- Solvent-based
- Water-based

Our markets
- Automotive
- Building and Construction
- Consumer Products
- Electronic and Assembly Materials
- Emulsion Polymers
- General Assembly
- Nonwovens and Hygiene
- Packaging
- Paper Converting
- Woodworking

Committed to our communities

- Supporting STEM education and youth leadership development
- Volunteers reach more than 30 countries with 6,000-plus hours of service annually
- Company and employee donations total more than $1.5 million globally each year.

GLUDAN
Deutschland GmbH

Am Hesterkamp 2
D-21514 Büchen
Phone +49 (0) 41 55-49 75-0
Fax +49 (0) 41 55-49 75 49
Email: gludan@gludan.de
www.gludan.com

Member of IVK

Company

Year of formation
1977

Size of workforce
28

Ownership structure
GmbH

Sales channels
Own sale force, and agents, have a look on
our map www.gludan.com for contacts

Contact partners
Management:
ks@gludan.de

Application technology and sales:
od@gludan.de
sb@gludan.de

Further information
www.gludan.com

Range of Products

Types of adhesives
Hot melt adhesives
Dispersion adhesives
Glutine glue
Pressure-sensitive adhesives

Raw materials
Additives
Fillers
Polymers
Starch

Equipment, plant and components
for conveying, mixing, metering and
for adhesive application
measuring and testing

For applications in the field of
Paper/packaging
Bookbinding/graphic design
Wood/furniture industry
Construction industry, including floors,
walls and ceilings
Textile industry
Adhesive tapes, labels
Hygiene
Household, recreation and office

Gößl + Pfaff GmbH

Münchener Straße 13
D-85123 Karlskron
Phone +49 8450-9320
Fax +49 8450-932 13
Email: info@goessel-pfaff.de
www.goessl-pfaff.de

Member of IVK

Company

Year of formation
1984

Size of workforce
22

Managing partners
Roland Gößl
Josef Pfaff

Sales channels
Technischer Vertrieb/Web-Shop

Contact partners
Management:
Johannes Pfaff

Application technology and sales:
Franziska Haller
Martina Reithmeier

Range of Products

Types of adhesives
Reactive adhesives

Equipment, plant and components
for conveying, mixing, metering and for
adhesive application

For applications in the field of
Wood/furniture industry
Construction industry, including floors, walls
and ceilings
Electronics
Mechanical engineering and equipment
construction
Automotive industry, aviation industry

Grünig KG

Häuserschlag 8
97688 Bad Kissingen
Phone +49 (0) 9736 7571-0
Fax +49 (0) 9736 7571-29
Email: info@gruenig-net.de
www.gruenig-net.de

Member of IVK

Company

Year of formation
1961

Size of workforce
31

Managing partners
Thomas Ulsamer
Sabine Ulsamer

Ownership structure
KG

Sales channels
Phone, FAX, Email, Post

Contact partners
Management:
Thomas Ulsamer

Application technology and sales:
Dietmar Itt
Andreas Schwab
Reinhold Teufel

Range of Products

Types of adhesives
Dispersion adhesives
Vegetable adhesives, dextrin and starch adhesives

Equipment, plant and components
for conveying, mixing, metering and for adhesive application

For applications in the field of
Paper/packaging
Bookbinding/graphic design
Wood/furniture industry
Hygiene

Fritz Häcker GmbH + Co. KG

Im Holzgarten 18
D-71665 Vaihingen/Enz
Phone +49 (0) 70 42-94 62-0
Fax +49 (0) 70 42-9 89 05
Email: info@haecker-gel.de
www.haecker-gel.de

Member of IVK

Company

Year of formation
1885

Size of workforce
22

Ownership structure
private owned company

Sales channels
Distributers worldwide

Contact partners
Management:
Ralf Müller

Range of Products

Types of adhesives
Gelatine based Adhesives

Types of sealants
Other

Raw materials
Technical Gelatine

Equipment, plant and components
for conveying, mixing, metering and for
adhesive application
measuring and testing

For applications in the field of
Paper/packaging
Bookbinding/graphic art
Tissue and towels
Box covering and lamination

Henkel
AG & Co. KGaA
Henkelstraße 67
D-40191 Düsseldorf
Phone +49 (0) 211-797-0
www.henkel.com

Member of IVK, FCIO, FKS, VLK

Company

Ownership structure
AG & Co. KGaA

Contact partners
Business Unit Adhesive Technologies
Headquarters Düsseldorf:
Phone +49 (0) 211-797-0
www.henkel-adhesives.com

Henkel Central
Eastern Europe GmbH
Phone +43 (1) 7 11 04-0
www.henkel.at
www.henkel-cee.com

Henkel & Cie. AG
Phone +41 (61) 825-70 00
www.henkel.ch

Henkel Belgium N.V.
Phone +32 (2) 421-27 11
www.henkel.be

Further information
Henkel Adhesive Technologies is leading
with high-impact solutions worldwide. We
lead today's and build tomorrow's markets
with our unique portfolio of breakthrough
innovations, tailor-made solutions and
strong brands in adhesives, sealants and
functional coatings. We combine innovation
and technology leadership and close
customer partnership to deliver solutions
that are an essential part of countless
industrial and consumer goods. We utilize
our global presence and our expert

Range of Products

Types of adhesives
Hot melt adhesives
Reactive adhesives
Solvent-based adhesives
Dispersion adhesives
Vegetable adhesives, dextrin and
starch adhesives
Pressure-sensitive adhesives
Instant adhesives
Structural adhesives
Anaerobic Adhesives
Laminating Adhesives

Types of sealants
Acrylic sealants
Butyl sealants
PUR sealants
Silicone sealants
MS/SMP sealants
Other

knowledge to offer a winning combination
of best-in-class service and leading
technologies to customers and consumers
around the world.

Our industrial product portfolio is organized
into five Technology Cluster Brands -
Loctite, Bonderite, Technomelt, Teroson
and Aquence. For consumer and professional markets, we focus on the four global
brand platforms Pritt, Loctite, Ceresit and
Pattex.

The future looks lighter

Electric mobility not only changes the way we drive – it also fundamentally impacts the construction of cars. The integration of heavy batteries with an average weight of 500 kilograms requires new concepts and materials for lightweight constructions. As a partner of the automotive industry, Henkel Adhesive Technologies offers a unique portfolio of high-impact solutions and technologies to enable the development of safer, more sustainable, and more efficient future cars.

More Sustainable

Adhesives are an integral part of modern lightweighting concepts. They enable the material mix of steel with aluminum and composites and help reduce a car's weight by up to 15 percent.

Safer

Innovative adhesive solutions ensure the optimal temperature management of lithium-ion batteries and prevent them from over-heating. The materials also enable compact constructions, which enhance the battery protection and improve safety in the event of a crash.

More efficient

New adhesive solutions in the engine, gearbox and electronics contribute to higher durability and operational safety. They also allow an optimized design of the powertrain system, as well as making electric cars lighter and more efficient.

 **IMCD**

IMCD N.V.
Wilhelminaplein 32
3072 DE Rotterdam, Netherlands
Phone +31 10 290 86 84
Fax +31 10 290 86 80
Email: coatings.construction@imcdgroup.com
www.imcdgroup.com/business-groups/
coatings-construction

Member of IVK

Company

Ownership structure
IMCD N.V. is a leading company in sales, marketing and distribution of specialty chemicals and food ingredients. With a revenue of € 2,379 million and 2,800 professionals in more than 50 countries, IMCD provides its partners with optimum tailored solutions for multi-territory distribution management in EMEA, Asia-Pacific and Americas.

Management
Piet van der Slikke, CEO IMCD N.V.
Frank Schneider, Director Business Group Coatings & Construction

Contact partners
Germany & International
Heinz J. Küppers, Market Manager Adhesives
Email: heinz.kueppers@imcd.de

Switzerland:
Yavuz Erol, Sales Manager
Email: yavuz.erol@imcd.ch
Armin Landolt, Sales & Marketing Assistant
Email: armin.landolt@imcd.ch

Austria and South East Europe:
Szabolcs Antal, Business Unit Manager
SEE - Coatings & Construction, Synthesis,
Lubricants & Fuels
Email: szabolcs.antal@imcd.at

Scandinavia:
Xoan Au, Business Unit Manager Coatings & Construction
Email: xoan.au@imcd.se

Range of Products

Types of adhesives
Solvent-based adhesives
Solvent-free adhesives
Dispersion adhesives
Pressure sensitive adhesives

Raw materials
Additives, Functional Fillers, Resins, Pigments, Solvents, Polymers

For applications in the field of
Paper/Packaging
Woodworking/Joinery
Building, Construction, Civil Engineering
Electronics
Automotive
Textile
Tapes, Labels, Graphic Arts

Services
Customer support
Guide formulations
Seminars

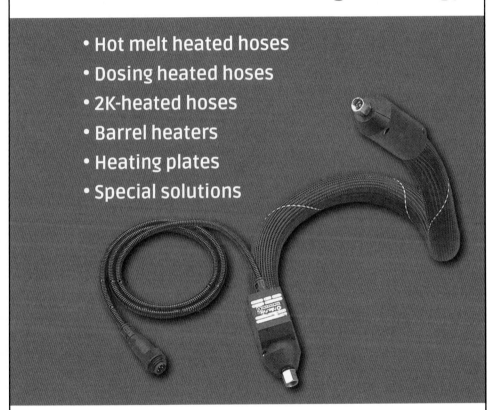

Klebstoffe

Jowat SE
Ernst-Hilker-Straße 10 – 14
D-32758 Detmold
Phone +49 (0) 52 31-7 49-0
Email: info@jowat.de
www.jowat.com

Member of IVK, FKS

Company

Year of formation
1919

Size of workforce
approx. 1,200

Managing partners
Klaus Kullmann
Ralf Nitschke
Dr. Christian Terfloth

Ownership structure
Shareholder company (not publicly traded)

Subsidiaries
23 worldwide

Sales channels
Own affiliations and distributors

Contact partners
Productmanagement
Ingo Horsthemke

Range of Products

Types of adhesives
Hot melt adhesives
Reactive adhesives
Solvent-based adhesives
Dispersion adhesives
Pressure-sensitive adhesives
Separating agents

For applications in the field of
Paper/packaging
Bookbinding/graphic design
Wood/furniture industry
Construction industry, including floors,
walls and ceilings
Electronics
Automotive industry, aviation industry
Textile industry
Adhesive tapes, labels
Upholstery, mattresses

Jowat | Your Partner in Bonding

**Our
word is
our bond.**
Since 1919

100 years of service – always on point.

Jowat supplies individual adhesive solutions for many industries.
Our commitment to quality does not end with the sale, though.
We also stand ready to assist you should you have any questions
about first use, applications or cleaning.
This is our superior service – made in Detmold.

www.jowat.com

Kaneka

KANEKA BELGIUM NV

The Dreamology Company
—Make your dreams come true—

Kaneka Belgium N. V.
MS Polymer Division
Nijverheidsstraat 16
B-2260 Westerlo (Oevel)
Telefon +32 (0) 14 - 25 45 20
E-Mail: info.mspolymer@kaneka.be
www.kaneka.be

Member of IVK

Company

Year of formation
Kaneka Belgium N.V. was founded in 1970 as the European production site of the globally acting Kaneka Corporation, Japan.

Further information
Kaneka products have conquered the European market, becoming a synonym for premium quality raw materials, with the brand MS POLYMER basically defining a new group of adhesives and sealants. Other brands of Kaneka's MS Polymer Division are SILYL and XMAP.

Production/Technical Service contact
Kaneka Belgium N.V.
MS Polymer Division
Nijverheidsstraat 16
B-2260 Westerlo (Oevel)
Phone +32 14-25 78 67
Email: info.mspolymer@kaneka.de

Marketing contact
Kaneka Belgium N.V.
MS Polymer Division
Nijverheidsstraat 16
B-2260 Westerlo (Oevel)
Phone +32 14-25 45 20
Email: info.mspolymer@kaneka.de

For contact in D, A, CH, CEE
Werner Hollbeck GmbH
Karl-Legien-Straße 7
D-45356 Essen
Phone +49 (0) 2 01-7 22 16 16
Fax +49 (0) 2 01-7 22 16 06
Email: info@hollbeck.de

Range of Products

Raw materials
MS Polymer, SILYL and XMAP are reactive high performance polymers based on polyether, or polyacrylate, respectively. Customers can select between moisture curing grades and radical cure types including UV-cure. The polymers' characteristics produce elastic adhesives and sealants, but also high strength types and coatings can be formulated.

End products' features
Solvent and isocyanate free 1-K/ 2-K-reactive adhesives and sealants
Pressure sensitive adhesives (PSA)
Oil resistance, temperature resistance (150 °C permanent use, XMAP)
Low gas and moisture permeability
High UV-resistance
Adhesive blends with epoxies

For applications in the field of
Wood/furniture industry
Construction industry, including floors, walls and ceilings
Electronics
Automotive industry, transportation industry
Adhesive tapes, labels
Household, recreation and office
General industry
Shipbuilding industry
Waterproofing and roof coatings

 **KEYSER & MACKAY**

Keyser & Mackay
German branch office
Industriestraße 163
D-50999 Köln (Rodenkirchen)
Phone +49 (0) 22 36-39 90-0
Fax +49 (0) 22 36-39 90-33
Email: info.de@keysermackay.com
www.keysermackay.com

Member of IVK

Company

Founded
1894

Size of workforce
120

Subsidiaries
Headquarter in The Netherlands

Subsidiaries in Germany, Belgium, France,
Switzerland, Poland, Spain

Contact partners
Mr. Robert Woizenko
Email: r.woizenko@keymac.com

Types of adhesives
Hot melt adhesives
Reactive adhesives
Solvent-based adhesives
Dispersion adhesives
Pressure-sensitive adhesives

Types of sealants
Acrylic sealants
Butyl sealants
Polysulfide sealants
PUR sealants
Silicone sealants
MS/SMP sealants

Range of Products

Raw materials
Resins: hydrogenated hydrocarbon resins,
C5/C9 hydrocarbon resins, pure monomer
resins, modified rosin resins, resin dispersions

Polymers: amorphous polyolefins (APO),
EVA, SIS/ SBS/ SEBS, PE / PP waxes,
FT waxes, STP, acrylic copolymers, acrylic
dispersions

Fillers: precipitated calcium carbonate, talc,
dolomite

Additives: oxazolidines, silanes, adhesion
promoters, flame retardants, polyols

For application in the field of
Paper/packaging
Bookbinding/graphic design
Wood/furniture industry
Construction industry, including floors, walls
and ceilings
Electronics
Mechanical engineering and equipment
construction
Automotive industry, aviation industry
Textile industry
Adhesive tapes, labels
Hygiene
Household, recreation and office

Kiesel Bauchemie GmbH u. Co. KG

Wolf-Hirth-Straße 2
D-73730 Esslingen
Phone +49 711 93134-0
Fax +49 711 93134-140
Email: kiesel@kiesel.com
www.kiesel.com

Member of GEV

Company

Year of formation
1959

Size of workforce
160

Ownership structure
family owned

Subsidiaries
Sales offices in Benelux, Poland, Czech Republic, Switzerland, France

Sales channels
wholesale to professional installers

Contact partners
Management & Sales:
Beatrice Kiesel-Luik

Application technology:
Ulrich Lauser

Range of Products

Types of adhesives
Reactive adhesives
SMP-Products
Silane-modified polymers
Dispersion adhesives
Cement based adhesives

Types of sealants
Acrylic sealants
PUR sealants
Cement based sealants
Dispersion sealants

For applications in the field of
Construction industry, including floors, walls and ceilings

Kisling
Deutschland GmbH

Bürgermeister-Seidl-Str. 2
D-82515 Wolfratshausen
E-Mail: info_de@kisling.com
Tel +49 8171 99982 30
Fax +49 322 224 299 35

Member of IVK, FKS

Company

Year of formation
2000

Size of workforce
7

Ownership structure
100 % subsidiary of Kisling AG, Switzerland

Sales channels
Direct sales, distributors

Contact partners
Managing Director:
Dr. Dirk Clemens

Head of Sales:
Frank Schöckel

Range of Products

Types of adhesives
Structural Adhesives
Anaerobic Adhesives
Cyanoacrylate Adhesives
RTV Silicones
Potting Compounds

For applications in the field of
Electrical Technology & Electronics
Mechanical & Plant Engineering
Pumps & Valves
Automotive Engineering
Consumer Goods
Maintenance and Repairs

KLEIBERIT Adhesives
KLEBCHEMIE M.G. Becker GmbH & Co. KG
Max-Becker-Straße 4
76356 Weingarten/Germany
Phone +49 7244 62-0
Fax +49 7244 700-0
Email: info@kleiberit.com
www.kleiberit.com

Member of IVK

Company

Year of formation
1948

Size of workforce
over 600 worldwide

Ownership structure
Owner Managed GmbH & Co. KG

Subsidiaries
Australia, France, USA, Canada, UK, Japan, China, Singapore, Russia, Brazil, India, Mexico, Ukraine, Turkey, Belarus

Sales channels
Direct, Wholesale

Contact partners
Management:
Dipl. Phys. Klaus Becker-Weimann
Dr. Achim Hübener
Sales:
Wolfgang Hormuth

Further information
Specialist in PUR-Adhesive-Technology
Competence PUR

Range of Products

Types of adhesives
PUR Hot melt
Reactive Hot melt (PUR, POR)
Holt melt (EVA, PO, PA)
1-C and 2-C reactive adhesives
(PUR, STP, Epoxy)
PUR foam systems
Dispersion adhesives
(Acrylat, EVA, PUR, PVAC)
Sealants and assembly adhesives
Pressure-sensitive adhesive
EPI-systems
Solvent-based adhesives

Coating systems
Kleiberit HotCoating® based on PUR
TopCoating based on UV lacquer

For applications in the field of
Wood- and furniture industry
Profile wrapping
Construction industry including floors, walls, ceilings and façade elements
Sandwich Panels
Textile industry
Automotive industry
Ship and boat building
Bookbinding industry
Surface finishing
Doors, windows, stairs and flooring
Filtration industry
Paper and packaging industry

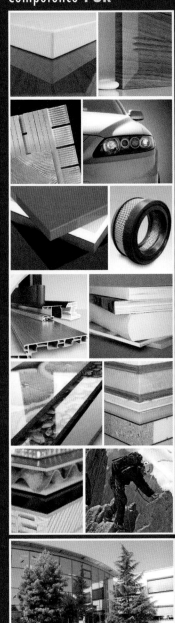

Kömmerling Chemische Fabrik GmbH
Zweibrücker Straße 200, D-66954 Pirmasens
Phone +49 (0) 63 31 56-20 00, Fax +49 (0) 63 31 56-19 99
Email: info@koe-chemie.de, www.koe-chemie.de
Member of IVK

Company

Year of formation
1897

Size of workforce
450

Ownership structure
H.B. Fuller

Subsidiaries
Kömmerling Chimie SARL, Strasbourg (F)
Kommerling UK LTD, Uxbridge (UK)
Kömmerling chemische Fabrik GmbH,
Beijing (China)

Sales channels
B2B, Trading

Further information
Kömmerling Chemische Fabrik GmbH is a
leading international manufacturer of high
quality adhesives and sealants. Established
over 115 years ago, Kömmerling is today a
major systems supplier for the Glass,
Transport, Construction, Industrial Assembly
and Renewable Energy industries. Kömmerling today is part of H.B. Fuller creating a
global Top 2 pure play adhesive company.

Range of Products

Types of adhesives
Hot melt adhesives
Reactive adhesives
Solvent-based adhesives
Dispersion adhesives
Pressure-sensitive adhesives

Types of sealants
Butyl sealants
Polysulfide sealants
PUR sealants
Silicone sealants
MS/hybrid sealants

For applications in the field of
Construction
Automotive
Marine
Coil coating
Shoe industry
Insulating glass
Photovoltaic, Solar Thermal
Window Bonding
Structural glazing
Wind energy

Krahn Chemie Deutschland GmbH
Grimm 10
D-20457 Hamburg
Phone +49 (0) 40-3 20 92-0
Fax +49 (0) 40-3 20 92-3 22
Email: info.de@krahn.eu
www.krahn.de

Member of IVK

Company

Krahn Chemie Deutschland is a speciality chemical distribution company and represents renowned, globally operating manufacturers in Europe. One of our core segments is the adhesives & sealants industry to which we provide raw materials.

Year of formation
1972

Size of workforce
170

Managing directors
Dr. Rolf Kuropka
Axel Sebbesse

Ownership structure
Otto Krahn (GmbH & Co.) KG founded 1909

Subsidiaries
Krahn Chemie Polska Sp. z o.o.
Krahn Chemie Benelux BV
Krahn France
Krahn Italia

Contact partners
Business Segment Manager
Adhesives & Sealants
Thorben Liebrecht
Email: thorben.liebrecht@krahn.eu

Application Technology and Sales:
Marcus Wedemann
Email: marcus.wedemann@krahn.eu

Supplying partners
BASF, Wanhua, Toyal, Valtris, Eastman, Exxon Mobil, Celanese, Chromaflo, Lanxess, Lord, Oxea, Tosoh, Dynaplak

Range of Products

Raw materials
Additives:
Pigments, Pigment Pastes
Silanes, Rheology Modifiers,
Plasticizers
Biocides

Dispersions:
Acrylic Dispersions
PU Dispersions
Polychloroprene Latex
Biobased Dispersions

Resins and Hardeners:
Saturated Polyester Resins
EH Diol, Isocyanates
Epoxy Hardeners

Polymers:
CR, CSM, EVA, PiB, PVC, PVB

Adhesion Promoters
Fillers

For applications in the field of
Paper/packaging
Bookbinding/graphic design
Wood/furniture industry
Construction industry, including floors, walls and ceilings
Electronics
Traffic Industry, Automotive industry, aviation industry
Textile industry
Adhesive tapes, labels
Mechanical engineering and equipment construction DIY

Kraton Polymers Nederland B.V.
Transistorstraat 16
NL-1322CE Almere
Phone +31 36 5462 800
Email: info@kraton.com

Member of IVK

Company

Contact details
info@kraton.com

Sustainable Solutions.
Endless Innovation.
At Kraton, we deliver exceptional value
to our customers through compelling,
innovative and sustainable solutions. We
create, develop and manufacture renewable
chemicals and specialty polymers to meet
our customers' performance and market
needs. We contribute to a sustainable future
by combining superior product performance
and quality, with reliability of global supply.
We live our core values every day, including taking ownership for doing what is safe,
honest and ethical. At Kraton, we seek
solutions that are sustainable, and believe
innovation is endless.

Range of Products

Types of adhesives
Hot melt adhesives
Solvent-based adhesives
Pressure-sensitive adhesives

Raw materials
Rosin Esters
Dispersions
Modified Rosin
AMS resins
AMS Phenolic resins
Styrenated Terpenes
Terpene phenolics
Polyterpenes
Hot-Melt Polyamides
SBS
SIS
SIBS
SEBS
SEPS

For applications in the field of
Packaging
Tapes
Labels
Building and construction / flooring
Hygiene
Bookbinding
Woodworking
Automotive
Aerospace

LANXESS Deutschland GmbH

Kennedyplatz 1
D-50569 Cologne
Phone +49 (0) 221-88 85- 0
www.lanxess.com

Member of IVK

Company

The global chemical company LANXESS with head office in Cologne, Germany, has been listed at the German stock exchange since 2005. More than 15,400 employees are working for LANXESS worldwide.

The product portfolio consists of 4 segments:
- Advanced Intermediates
- Specialty Additives
- Engineering Materials
- Performance Chemicals
The turnover was 7.2 bn € in 2018.

Range of Products

Biocides
Wide range of biocides under the brand names Preventol® and Metasol®:
- In-can preservatives for all kinds of water-based adhesives
- Fungicides for film protection of sealants, adhesives and grout fillings
- Consultation on microbiological questions

Contact – for biocides
Dr. Peter Wachtler
Material Protection Products
Technical Marketing – Industrial Preservation
Phone: +49 (0) 221 8885 4802
Email: peter.wachtler@lanxess.com

 **L&L Products**

L&L Products Europe
Am Sundheimer Fort 1-11
D-77694 Kehl
Phone 07851-6299-70
E-Mail: info.europe@llproducts.com
www.llproducts.com

Member of IVK

Company

Year of formation
1958 in Romeo USA –
2002 foundation of the GmbH

Size of workforce
1.200 worldwide

Managing partners
L&L Products Holding Inc.

Nominal capital
25.000 €

Ownership structure
Familys Lane & Ligon

Sales channels
Direct and Distribution

Contact partners
Management:
Matthias Fuchs

Application technology and sales:
Jean-Michel Hollaender

Range of Products

Types of adhesives
Hot melt adhesives
Reactive adhesives
Pressure-sensitive adhesives

Types of sealants
Acrylat sealants
PUR-sealants
MS/SMP sealants
Other

For applications in the field of
Electronics
Mechanical engineering and equipment
construction
Automotive industry, aviation industry

Bondexpo

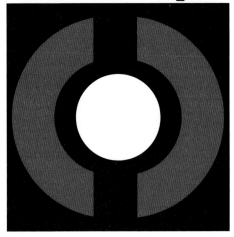

International trade fair for bonding technology

📅 **07.–10. OCTOBER 2019** 📍 **STUTTGART**

Connecting with the Best Technologies

The 13th Bondexpo will present the most up-to-date worldwide offerings covering technologies and processes for bonding, joining and fastening in industrial production and integrated assembly.

- Raw materials for adhesives and sealants
- Machines, systems and accessories for the adhesives manufacturing inductry
- Adhesives and sealants
- Machines, systems and accessories for the adhesives processing industry
- Sealing-, measuring and test technology

🌐 **www.bondexpo-messe.com**

Lohmann GmbH & Co. KG
Irlicher Straße 55
D-56567 Neuwied
Phone +49 (0) 26 31-34-0
Fax +49 (0) 26 31-34-66 61
Email: info@lohmann-tapes.com
www.lohmann-tapes.com

Member of IVK

Company

Year of formation
1851

Size of workforce
approx. 1,800 worldwide

Managing directors:
Elmar Boeke
Martin Schilcher

Subsidiaries
I, F, E, PL, A, GB, NL, DK, SE, RU, UA, USA, China, Korea, Singapore, India, Mexico and Turkey

Sales channels
Market segments:
Graphics, Consumer Goods, Building & Construction, Transportation, Mobile Communication, Medical Technology, Renewable Energies, Electronics and Hygiene

Contact partners
Head of PR and Corporate Communication
Christina Barg-Becker M. A.

Ralph Uenver, Head of Corporate and Marketing Communication

Further information
Lohmann offers mainly customized bonding solutions and takes care of its customers from the first idea up to automatic applications.
This is also shown in the company logo: „The Bonding Engineers".

Range of Products

Types of adhesives
Hot melt adhesives
Reactive adhesives
Solvent-based adhesives
Dispersion adhesives
Pressure-sensitive adhesives

Types of sealants
Acrylic sealants

Raw materials
Polymers

Equipment, plant and components for
Conveying, mixing, metering
Adhesive application
Surface pretreatment
Adhesive curing and drying
Measuring and testing

For applications in the field of
Paper/packaging
Plate mountig tapes
Wood/furniture industry
Building & Construction industry, including floors, walls and ceilings, structural glazing
Electronics
Automotive industry
Adhesive tapes, die-cuts, labels
Mobile Communication
Renewable Energies
Medical Technology/diagnostics
Hygiene
Consumer Goods
High security cards

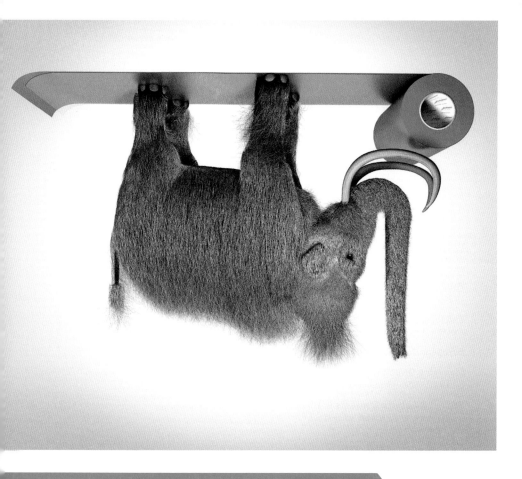

A clear case for LOHMANN

When it comes to unusual challenges:
we will bond it for you.

Uniting the amazing abilities of structural bonding with the easy handling of regular bonding tape?
Yes, we can do that for you:
Our DuploTEC® SBF (Structural Bonding Films) range offers a powerful alternative to conventional
bonding solutions. It is just one example of how we cover the entire value chain including cutting,
slitting and die cutting and coating. We provide customer-focused solutions for many industries
including Home Appliances & Electronics, Automotive, Building and Construction.

The Bonding Engineers

www.lohmann-tapes.com

LORD Germany GmbH

Itterpark 8
D-40724 Hilden
Phone +49 (0) 21 03-2 5 23 10
Fax +49 (0) 21 03-2 52 31 97
Email: monika_schulz@lord.com
www.lord.com/emea

Member of IVK

Company

Founded in
1924 by Hugh C. Lord

Employees
3,000 (LORD Corp. worldwide)

European Technical Center Hilden

Managing Directors
Dr. Dirk L. Schröder, Vincent Javerzac

Technical Service Contacts in Hilden
LORD Rubber to Metal adhesives
Dr. René Wambach
Email: rene_wambach@lord.com

LORD Automotive OEM 2K adhesives
Dipl. Ing. Marcus Lämmer
Email: marcus_laemmer@lord.com

LORD Industrial adhesives
Dipl. Ing. Marcus Lämmer
Email: marcus_laemmer@lord.com

LORD Electronic materials
Dipl. Ing. Marcus Lämmer
Email: marcus_laemmer@lord.com

LORD Flock adhesives and rubber coatings
Dr. Christiane Stingel
Email: christiane_stingel@lord.com

Additional Information
LORD Corp. is the global leader in producing
rubber to metal adhesives used in a wide
variety of automotive, aerospace and industrial
applications. In addition LORD is producing
solvent based and aqueous flock adhesives and
coatings for elastomeric components. On the
structural assembly part LORD is producing 2K
metal and composite structural adhesives

Range of Products

Type of Adhesives
2K Structural Adhesives (LORD®, FUSOR®,
VERSILOK®)
Solvent based rubber to metal adhesives
(CHEMLOK®, CHEMOSIL®)
Solvent based reactive 1K adhesives
(LORD®, CHEMLOK®, FLOCKSIL®)
Water based reactive 1K adhesives
(CUVERTIN®, SIPIOL®)

For applications
Electronics: potting, coating, waver adhesives
Industrial: metal, composite, plastic
Elastomeric/Rubber: Rubber to metal, flock,
slip-coating, anti-squeeze
Automotive (OEM): 2K cold cure hem flange
adhesives, 2K LED cure hem flange sealer
(body shop applied), 2K repair adhesives,
vibration mounts, magnet rheological fluids
(MR fluids) and MR devices
Aerospace: 2K OEM structural adhesives,
2K repair adhesives, coatings, vibration mounts,
active vibration control devices

which are use e.g. in automotive hem flange
bonding. Another focus is on electronic
adhesives used for potting and encapsulation
as well as electrically conductive adhesives
and grease for thermal management.

The European Customer Service Center in
Hilden is the research and development center
for customized solutions as well as the main
European training center.

LUGATO
GmbH & Co. KG

Großer Kamp 1
D-22885 Barsbüttel
Phone +49 (0) 40-6 94 07-0
Fax +49 (0) 40-6 94 07-1 10
Email: info@lugato.de
www.lugato.de

Member of IVK, FCIO,
Deutsche Bauchemie e.V.

Company

Year of formation
1919

Size of workforce
About 135

Ownership structure
GmbH & Co. KG, associated company of the
Ardex GmbH since 2005

Sales channels
DIY stores

Contact partners
Application technology and sales:
info@lugato.de

Further information
www.lugato.de
www.lugato.com

Range of Products

Types of adhesives
Dispersion adhesives
Cementitious adhesives
MS/SMP adhesives

Types of sealants
Acrylic sealants
Silicone sealants
MS/SMP sealants

For applications in the field of
Household, recreation and office

Mapei Austria GmbH

Fräuleinmühle 2
A-3134 Nußdorf o.d. Traisen
Phone +43 (0) 27 83-88 91
Fax +43 (0) 27 83-88 91-1 25
Email: office@mapei.at
www.mapei.at

Member of IVK, FCIO

Company

Year of formation
1980

Size of workforce
140 employees

Managing partners
Mapei SpA, Milano, Italien

Subsidiaries
Mapei Kft, Ungarn, Mapei sro,
Tschechische Republik;
Mapefin Austria GmbH

Contact partners
Management:
Mag. Andreas Wolf

Sales management:
Paul Solczykiewicz

Product management:
Stefan Schallerbauer

Range of Products

Types of adhesives
Hot melt adhesives
Reactive adhesives
Solvent-based adhesives
Dispersion adhesives

Types of sealants
Acrylic sealants
Silicone sealants

For applications in the field of
Construction industry, including floors,
walls and ceilings

MAPEI S.p.A.

MAPEI GmbH
IHP Nord – Bürogebäude 1
Babenhäuser Straße 50
D-63762 Großostheim

Via Cafiero 22
20158 Milano, Italy
Phone +39 02/376 731
Fax + 39 02/376 732 14
Email: mapei@mapei.it
www.mapei.com

Member of IVK

Company

Year of formation
1937

Size of workforce
more than 10,500

Managing partners
MAPEI S.p.A., Milano, Italy

Ownership structure
family-owned enterprise

Subsidiaries
87 subsidiaries in 35 countries

Sales channels
wholesalers

Contact partners
Management:
Flavio Terruzzi

Range of Products

Types of adhesives
Reactive adhesives
Dispersion adhesives
Pressure-sensitive adhesives

Types of sealants
Acrylics sealants
PUR sealants
Silicone sealants
MS/SMP sealants
Other

For applications in the field of
Construction industry, including floors,
walls and ceilings

merz+benteli ag
more than bonding

Merbenit Gomastit Merbenature

merz+benteli ag
Freiburgstrasse 616
CH-3172 Niederwangen
Phone +41 (31) 980 48 48
Fax +41 (31) 980 48 49
Email: info@merz-benteli.ch
www.merz-benteli.ch

Member of FKS

Company

Organisation
merz+benteli ag was founded in 1918 to supply the Swiss watch industry with innovative adhesives. The enterprise is a family-owned joint-stock company specialised in adhesives and sealants. R&D and production in Switzerland.

Contact
Simon Bienz
Director Marketing & Sales

Distribution partners
DE/AT: Reiss Kraft GmbH
www.reiss-kraft.de / +49 7253 93 47 65

GR: Dialinas AE
www.dialinas.gr / +30 210 27 13 333

ES, PT, GB, IE, RU, DK, TR, SE, NO, FI, IT, BE, NL, LU, FR, CZ, SI
Distribution Groupe Europe
www.dge-europe.com / +31 172 436 361

HU: Güteber Kft.
probond@berenyizoltan.hu / +36 1 213 5005

SL: Koop KOOP Trgovina d.o.o.
www.koop.si / +386 (7) 477-8820

IL: Rotal Adhesives & Chemicals Ltd.
www.rotal.com / +972 9 766 7990

USA, CAN, MEX: Chenso Inc.
www.chenso.com / +1 336 681 4131

Asia/Pacific: Innosolv Pty.
rick@innosolv.com / +61 (6) 9846 7871

Range of Products

Range of Products
SMP adhesives and sealants branded as GOMASTIT and MERBENIT.
Bio-based SMP sealants MERBENATURE
Patented MERBENTECH technology

Range of application
Elastic sealing for building construction:
Facades
Interior construction
Floor
Sanitary
Glazing
Roof
Fire protection

Elastic adhesives for industrial constructions:
Industry
Automotive
Marine

Minova CarboTech GmbH
Bamlerstraße 5 d
D-45141 Essen
Phone +49 (0) 201 80983 500
Fax +49 (0) 201 80983 9605
Email: info.de@minovaglobal.com
www.minovaglobal.com

Member of IVK

Company

Range of Products

Holding company
Orica Ltd, 1st Nicholson Street,
3000 Melbourne, NSW, Australien

ORICA Subsidiaries worldwide
Australia, Chile, China, Germany, France,
Great Britain, India, Italy, Canada,
Kazakhstan, Austria, Poland, Romania,
Russia, Sweden, Switzerland, Singapore,
Spain, South Africa, Taiwan,
Czech Republic, Turkey, Ukraine, USA

Contact partners
Management:
Michael J. Napoletano, Andreas Humann

Application technology and sales
Paul van der Lubbe
Phone +49 (0) 201 80983 500
Mobil +43 (0) 43 664 8459317
Email: paul.vanderlubbe@minovaglobal.com

Further information
Minova has more than 40 years of worldwide experience in the fields of water
control, ground stabilisation, anchoring,
rock bolting and flooring adhesives.

We are a leading supplier of products and
services for mining, tunnelling, geotechni-
cal engineering, special underground civil
engineering and floor technology.

Types of adhesives
Reactive adhesives

For applications in the field of
Wood/furniture industry,
Construction industry, including floors,
walls and ceilings

We supply an extensive range of products for:

- Anchoring systems for mining and tunnelling as well as special underground civil
  engineering
- Resin systems for sealing, rock stabilisation and reinforcement
- Accessories for injection and anchoring
  systems
- Adhesives for various floors and floor
  coverings
- Resin systems and accessories for
  sewerand ductrepair

MÖLLER CHEMIE
CHEMICAL PARTNERSHIP SINCE 1920

Möller Chemie GmbH & Co.KG
Bürgerkamp 1
D-48565 Steinfurt
Phone +49 (0) 2551 9340-0
Fax +49 (0) 2551 9340-60
Email: info@moellerchemie.com
www.moellerchemie.com

Member of IVK

Company

Year of formation
Company was founded 1920

Size of workforce
102

Nominal capital
Turnover ~ 72 Mill. €

Ownership structure
Family owned; Private

Sales channels
EU direct or via agent in UK, Balkan area

Contact partners
Management:
R. Berghaus

Application technology and sales:
U. Banseberg, S. Bruns

Further information
Specialized in Additive - Defoamer, Wetting and Dispersing agent, UV-light Stabilizer, Silicone surfactants, Rheology agents.
Epoxy curing agent (solvent- and water based) of DDCHEM; Reactive diluents, MXDA, Solvents, Silanes. Hydrophobic and Hydrophilic fumed Silica of Orisil.
Alpha Olefine of Chevron Phillips (AlphaPlus). Isoparaffine, Plasticizer, Isotridecanol, etc. Amine and Metal catalysts, Crosslinker, Other.

Range of Products

Raw materials
Additives
Fillers
Resins
Solvents
Polymers
Starch

Equipment, plant and components
for conveying, mixing, metering and for adhesive application

For applications in the field of
Paper/packaging
Bookbinding/graphic design
Wood/furniture industry
Construction industry, including floors, walls and ceilings
Electronics
Mechanical engineering and equipment construction
Automotive industry, aviation industry
Textile industry
Adhesive tapes, labels
Hygiene
Household, recreation and office

MÜNZING
CREATING ADDITIVE VALUE

MÜNZING CHEMIE GmbH
Münzingstraße 2
D-74232 Abstatt
Phone +49 (0) 71 31-987-0
Fax +49 (0) 71 31-987-202
Email: sales.pca@munzing.com
www.munzing.com

Company

Year of formation
1830

Size of workforce
400

Managing partners
family owned

Subsidiaries
MÜNZING North America,
Bloomfield, NJ, USA
MÜNZING CHEMIE Iberia S.A.U.,
Barcelona, Spain
MÜNZING International S.a.r.l., Luxembourg
MÜNZING Micro Technologies GmbH, Elster-
aue, Germany
MÜNZING Shanghai Co. Ltd, P.R. China
MAGRABAR LLC, Morton Grove, IL, USA
MÜNZING Mumbai Pvt. Ltd., Mumbai, India
MÜNZING Australia Pty. Ltd., Somersby, NSW,
Australia
MÜNZING Malaysia SDN BHD, Sungai Petani,
Malaysia
MUNZING DO BRASIL, Curitiba/PR, Brazil
Süddeutsche Emulsions-Chemie GmbH,
Mannheim, Germany

Sales channels
direct and via distributors

Contact partners
Application technology:
Peter Bissinger
Tel. No. +49 (0) 71 31-987-174
Email: p.bissinger@munzing.com

Sales:
Dr. Nicholas Büthe
Tel. No. +49 (0) 71 31-987-148
Email: sales.pca@munzing.com

Range of Products

Raw materials
Additives: Defoamers, dispersants, rheology
modifiers, wetting and levelling agents

For applications in the field of
Paper/packaging
Bookbinding/graphic design
Wood/furniture industry
Construction industry, including floors, walls
and ceilings
Electronics
Mechanical engineering and equipment
construction
Automotive industry, aviation industry
Textile industry
Adhesive tapes, labels
Household, recreation and office

MUREXIN GmbH

Franz v. Furtenbach Straße 1
A-2700 Wr. Neustadt
Phone +43 (0) 26 22/2 74 01
Email info@murexin.com
www.murexin.com

Member of IVK, FCIO

Company

Year of formation
1931

Size of workforce
400

Ownership structure
Schmid Industrie Holding

Subsidiaries
Croatia, Romania, Czech. Republik,
Hungary, Slovenia,
Slovak. Republic, Poland, France

Contact partners
Management:
Bernhard Mucherl

Range of Products

Types of adhesives
Solvent-based adhesives
Dispersion adhesives
MS adhesives

Types of sealants
Bitumen sealings
Bitumen-free sealings
Joint sealings

For applications in the field of
Construction industry, including floors,
walls and ceilings

Nordmann, Rassmann GmbH
Kajen 2
D-20459 Hamburg
Phone +49 (0) 40 36 87-0
Fax +49 (0) 40 36 87-2 49
Email: info@nordmann.global
Internet: www.nordmann.global

Member of IVK

Company

Year of formation
1912

Size of workforce
450 employees

Managing Board
Dr. Gerd Bergmann, Carsten Güntner, Felix Kruse

Ownership structure
Family-owned

Subsidiaries
Austria, Bulgaria, Czech Republic, France, Hungary, India, Italy, Japan, Poland, Portugal, Romania, Serbia, Singapore, Slovakia, Slovenia, South Korea, Spain, Sweden, Switzerland, Turkey, United Kingdom, USA

Sales channels
Face-to-face sales, telephone, email, fax

Contact partners
Management: Henning Schild
Phone +49 (0) 40 36 87-248, Fax +49 (0) 40 36 87-72 48
Email: henning.schild@nordmann.global

Application technology and sales: Michael Herrmann
Phone +49 (0) 40 36 87-463, Fax +49 (0) 40 36 87-74 63
Email: michael.herrmann@nordmann.global

Further information
A leading multinational chemical distribution company with subsidiaries in Europe, Asia and North America, Nordmann distributes natural and chemical raw materials, additives and specialty chemicals around the world. As a sales and marketing organization, Nordmann links suppliers from across the globe to customers throughout the manu-facturing industry. Our success at Nordmann comes from the direct connections we build with customers from 26 different locations worldwide and through the extensive portfolio of high-quality, technically sophisticated raw materials and specialty chemicals our company provides. With the individualized solutions and services we offer, Nordmann helps manufacturers from all major industries produce their goods safely and efficiently.

We offer partners:
- an innovative product portfolio
- technical know-how and application-related expertise in top industries
- secure and reliable supply-chain solutions
- help with product development
- expertise and support concerning regulatory compliance around the world

Range of Products

Raw materials
Additives: antioxidants/stabilizers, cellulose ethers, dispersion powders, defoamers, PVA, polyethylene oxide, rheology modifiers, polyolefin waxes and copolymers (PE/PP), pigments, impact modifiers, flame retardants and synergists, carbon fibers, functional fillers, starch ethers, surfactants, thickeners

Polymers: styrenic block copolymers (SBS, SEBS, SEP, SIS, SIBS), chloroprene rubber (CR), PVDC

Resins: hydrocarbon resins, gum rosin and rosin ester derivates epoxy resins and hardeners, phenolic resins, alkyd resins

Reactive components: polyetherpolyols, polyester-polyols, PTMEG, polycaprolactones, isocyanates (MDI), catalysts, reactive diluents, chain extenders, monomers (methacrylates, hydroxymethacrylates, ethermeth-acrylates, aminomethacrylates)

Plasticizers: mineral oil based process oils, gas to-liquid (GtL) based process oils, DOTP, DPHP

Dispersions: styrene-acrylate, styrene-butadiene, vinyl-acetate, chloroprene, gum rosin ester, Polyurethane

For applications in the field of
Adhesive tapes, labels
Automotive industry, aviation industry
Bookbinding/graphic design
Construction industry, including floors, walls and ceilings
Electronics
Household, recreation and office
Hygiene
Mechanical engineering and equipment construction
Paper/packaging
Textile industry
Wood/furniture industry

Nordmann has been an independent family business since its foundation in 1912 and belongs to Georg Nordmann Holding AG. The company staffs a total of 450 employees and achieved € 460 million in turn-over in 2018. Based in Hamburg, Germany, Nordmann, Rassmann GmbH is the headquarter of Nordmann. For more details please refer to our website at www.nordmann.global.

Nynas GmbH
Marktplatz 2
D-40764 Langenfeld
Email: thorsten.wolff@nynas.com
www.nynas.com

Member of IVK

Company

Year of formation
1983

Size of workforce
10

Managing partners
Nynas AB Stockholm

Nominal capital
76,693

Subsidiaries
30 offices worldwide

Contact partners
Sales Manager Germany:
Thorsten Wolff

Application technology and sales:
Nina Lulsdorf
Jan-Peter Laabs

Range of Products

Process oils for adhesives/hotmelts

For applications in the field of
Paper/packaging
Construction industry, including floors, walls
and ceilings
Automotive industry, aviation industry
Textile industry
Adhesive tapes, labels
Hygiene
Household, recreation and office

Naphthenic oils for better adhesion

Nynas oils are widely used as plasticizers in adhesives, including hot melt adhesives for nonwovens, tapes and labels, and flooring adhesives. Their broad polymer and resin compability leads to higher peel adhesion, while enabling higher raw material flexibility. Let our technical experts help you achieve the best possible performance while meeting complex regulations – wherever you do business.

www.nynas.com/adhesive-plasticizers

Omnicol nv

Nijverheidsstraat 14
2380 Weelde (Belgium)
Phone +32 (0) 14 65 62 85
Email: info@omnicol.eu
www.omnicol.eu

Member of IVK

Company

Year of formation
1956

Size of workforce
76

Further information
Strong family ties
Omnicol has been a family business ever since it was established in 1956. This has many advantages. For example, values which are often ignored today are firmly embedded in the company. These include keeping promises and maintaining a strong customer focus. Omnicol is regarded as a reliable and solution-oriented partner. The head office is in Weelde (Belgium), which is also home to the production and R&D departments. From the sales offices in Ham (Belgium) and Hedel (the Netherlands), Omnicol products make their way to clients. The flat organisational structure guarantees short lines of communication and rapid answers.

Core business
Omnicol specialises in professional solutions for bonding construction materials. Bathrooms, swimming pools and hospitals, among many other projects, are professionally finished using Omnicol products as the connecting link between the bare substrate and the aesthetic finish. Omnicol also supplies powerful solutions for bonding bricks for external walls.

Range of Products

Types of adhesives
Reactive adhesives
Dispersion adhesives

Types of sealants
Silicone sealants

Raw materials
Additives

Equipment, plant and components
for conveying, mixing, metering and for adhesive application
for surface pretreatment
measuring and testing

For applications in the field of
Construction industry, including floors, walls and ceilings

The result: strong constructions with minimal joints. The streamlined image is emphasised by the variety of colour mixtures which are available. Omnicol is always innovating. As a result, Omnicol products are also used in the prefab industry.

Omya GmbH

Poßmoorweg 2
D-22301 Hamburg
Phone +49 (0) 221-37 75-0
Fax +49 (0) 221-37 75-390
Email: building.de@omya.com
www.omya.de

Member of IVK

Company

Contact
Gabriele Bender
gabriele.bender@omya.com

Range of Products

Raw materials
Additives:
Dispersants for water borne systems
Rheology Modifier (PU-/Acrylic thickeners,
Bentonites, Sepiolites)
Super Plasticizer (Dry Polycarboxylate Ether)

Industrial Minerals:
Calcium Carbonates, Dolomite,
Ultrafine PCC
Kaolin, Baryte
Flame Retardants (ATH)
Lightweight Fillers

Polymers:
Vinyl Acetate (VAC)
Acrylic and Styrene-Acrylic Copolymer
Alkyd and Polyester Resins
Styrene Butadiene Rubber (SBR)
Epoxy Resins, Hardeners and Reactive
Diluents
Hot Melt Resins (PA)

For applications in the field of
Paper/packaging
Bookbinding/graphic design
Wood/furniture industry
transportation-/automotive-/aerospace
industry
Construction industry, including floors,
walls and ceilings
Sealants
Casting-, potting materials
Textile industry
Household, recreation and office

Organik Kimya Netherlands B. V.

Chemieweg 7, Havenummer 4206
NL-3197KC Rotterdam Botlek,
Phone +31 10 295 48 20
Fax +31 10 295 48 29
Email: organik@organikkimya.com
www.organikkimya.com

Member of IVK

Company

Year of formation
1924

Size of workforce
500

Managing partners
Simone Kaslowski
Stefano Kaslowski

Nominal capital
100 %

Ownership structure
100 % familiy owned business

Subsidiaries
Distribution worldwide, production sites in
the Netherlands and Turkey

Sales channels
Direct sales and through distributors
worldwide

Contact partners
Management:
Stefano Kaslowski, General Manager

Application technology and sales:
Oguz Kocak, Sales Manager
Phone: +49 (0) 173-6 52 22 59
Email: o_kocak@organikkimya.com

Range of Products

Types of adhesives
Dispersion adhesives
Pressure-sensitive adhesives

Types of sealants
Acrylic sealants

Raw materials
Polymers

For applications in the field of
Paper/packaging
Bookbinding/graphic design
Wood/furniture industry
Construction industry, including floors,
walls and ceilings
Automotive industry, aviation industry
Textile industry
Adhesive tapes, labels

Sealants • Adhesives

Hermann Otto GmbH
Krankenhausstraße 14
D-83413 Fridolfing
Phone +49 (0) 86 84-9 08-0
Fax +49 (0) 86 84-9 08-5 39
Email: info@otto-chemie.com
www.otto-chemie.com

Member of IVK

Company

Management board
Johann Hafner
Matthias Nath

Size of workforce
450 employees

Contact
Technical Departement
Phone +49 (0) 8 68 49 08-4 60
Email: industry@otto-chemie.de

Sales manager industrial applications:
Marc Wüst
Phone +49 (0) 8 68 49 08-5 21
Email: marc.wuest@otto-chemie.de

Range of Products

Types of sealants and adhesives
1-comp. and 2-comp. silicones
1-comp. and 2-comp. polyurethanes
1-comp. and 2-comp. hybrids
acrylates

For applications in the field of
Renewable energies
Domestic appliances industry and
professional cooking technology
Lighting and electronics components
Light construction and composite elements
Laminating an coating
Heating, ventilation and plant construction

Panacol-Elosol GmbH
Member of Hönle Group
Daimlerstraße 8
D-61449 Steinbach/Taunus
Phone +49 (0) 61 71 62 02-0
Fax +49 (0) 61 71 62 02-5 90
Email: info@panacol.de
www.panacol.com

Member of IVK

Company

Year of formation
1978

Size of workforce
50

Managing Director
Florian Eulenhöfer

Nominal capital
255,645 €

Ownership structure
Panacol-Elosol GmbH is the German subsidiary of Panacol AG in Switzerland. Panacol Group is the subsidiary of Dr. Hönle AG

Sales channels
Sales team for Germany
International distribution network with worldwide sales partners

Contact partners
Sales Director and Application Engineering:
Dr. Detlef Heindl

Further information
Panacol is a leading international manufacturer of industrial adhesives as well as medical grade adhesives. In addition to Eleco-EFD, an affiliated company in France, Panacol provides an international network of distributors, which ensure a personal advisory service around the world. Since January 2008 Panacol has been a member of the Hönle Group and is benefiting from the numerous synergies in the field of industrial UV technology.

Range of Products

Types of adhesives
Reactive adhesives
Anaerobic adhesives
Cyanoacrylates
LED/UVA/visible light curing epoxy and acrylate adhesives
High temperature adhesives
Conductive adhesives
Medical grade adhesives
Structural adhesives

Types of sealants
Epoxies
Acrylics

Equipment
UV and LED equipment for adhesive curing from Hönle UV technology

Market Segments
Electronics/Applications on PCBs
Conformal coatings
Smart card/Die attach
Optics and fibre optics/Active alignment
Optoelectronics
Medical device assembly
Display bonding/Liquid gaskets
Loudspeaker assembly
Appliances
Automotive and Aviation Industry

experience. performance.

Paramelt B.V.
Costerstraat 18
NL-1704 RJ Heerhugowaard
The Netherlands
Phone +31 (0) 72 5 75 06 00
Email: info@paramelt.com
www.paramelt.com

Member of VLK

Company

Year of formation
1898

Size of workforce
522

Ownership structure
privately owned

Subsidiaries
Paramelt Veendam B.V.; Paramelt USA Inc.;
Paramelt Specialty Materials (Suzhou) Co., Ltd.

Sales channels
Sales offices in Sweden, Germany,
Netherlands, UK, France and Portugal and
a network of specialised distributors

Contact partners
Flexible Packaging: Leon Krings
Packaging & Labelling: Kristof Andrzejewski
Construction & Assembly: Wim van Praag

About Paramelt
Paramelt (established in 1898) has grown over
the years to become the leading global specialist in wax based materials and adhesives.
Today Paramelt operates from 8 production
locations around the globe in The Netherlands, USA and China. The company functions
through a series of global business units providing a structured approach to the key market
sectors in which we operate like e.g. packaging
and construction & assembly. For these markets we offer a comprehensive range of waxes,
water based adhesives, hot melt (pressure
sensitive) adhesives, water-based functional
coatings, solvent-based and PU adhesives.
Serviced by both regional sales offices, as well

Range of Products

Types of adhesives
Hot melt adhesives
Reactive adhesives
Solvent-based adhesives
Dispersion adhesives
Vegetable adhesives, casein, dextrin and
starch adhesives
Pressure-sensitive adhesives
Heat seal coatings

Types of sealants
Other

For applications in the field of
Paper/(Flexible) Packaging/Labelling
Construction industry/General Assembly

---

as a comprehensive network of distribution
partners, our customers can be assured of the
highest levels of local service and support.
Paramelt possess extensive experience in the
design and development of adhesives and
functional coatings to meet critical machine
and application requirements. The company
has built significant knowledge of performance
aspects needed to make our products effective
at all stages of the supply chain. Our products
are backed up by regional laboratories providing
comprehensive application and analytical testing facilities to ensure selection of the most
appropriate adhesive for your application.
Built on a tradition of partnership and trust;
underpinned by detailed knowledge gained
over more than 120 years of operation, Paramelt can offer real benefits to your operation.

PCC Specialties GmbH

Suedstrasse 13
47475 Kamp-Lintfort, Germany
Phone: +49 (0) 2842 12187-0
Fax: +49 (0) 2842 12187-299
Email: specialties@pcc.eu
www.pcc-specialties.eu

Member of IVK

Company

Founding year
2018

Size of workforce
14 employees

Ownership
Company of the PCC group

Distribution channel
International – direct

Contact
Management:
Dr. Uwe Zakrzewski

Application technology and sales:
Dorothea Müschenborn
Phone: +49 (0) 2842 12187-101
Email: dorothea.mueschenborn@pcc.eu

Further information
True to the motto "Making future technology applicable", PCC Specialties GmbH
is a problem solver, niche supplier and
innovation partner in the adhesives industry.
Together with our customers we address
the market challenges and develop tailor-
made products and formulations for your
individual requirements.

Range of Products

Types of adhesives
Hot-melt adhesives
Solvent-based adhesives
Dispersion adhesives
Pressure-sensitive adhesives

Raw materials
Additives
Resins
Polymers

For applications in the field of
Paper/Packaging
Wood/furniture industry
Construction incl. floor, wall and ceiling
Mechanical and apparatus engineering
Automotive, aviation industry
Textile industry
Adhesive tapes, labels
Household, hobby and office

PCI Augsburg GmbH

Piccardstraße 11
D-86159 Augsburg
Phone +49 (0) 8 21 59 01-0
Fax +49 (0) 8 21 59 01-3 72
Email: pci-info@basf.com
www.pci-augsburg.com

Member of IVK, FCIO

Company

Year of formation
1950

Size of workforce
800

Ownership structure
PCI Augsburg GmbH is part of BASF –
The Chemical Company

Subsidiaries
please refer to website for details

Sales channels
indirect/through distributors

Contact partners
Management:
please refer to website

Application technology and sales:
please refer to website

Further information
please refer to website

Range of Products

Types of adhesives
Reactive adhesives
Dispersion adhesives

Types of sealants
Acrylic sealants
PUR sealants
Silicone sealants

Equipment, plant and components
or conveying, mixing, metering and
for adhesive application

For applications in the field of
Construction industry, including floors,
walls and ceilings

PLANATOL®
smart gluing

Planatol GmbH
Fabrikstraße 30 - 32
D-83101 Rohrdorf
Phone +49 (0) 80 31- 7 20 - 0
Fax +49 (0) 80 31- 7 20 -1 80
Email: info@planatol.de
www.planatol.de

Niederlassung Herford:
Hohe Warth 15 - 21
D-32052 Herford
Tel.: +49 (0) 52 21 - 77 01-0
Fax: +49 (0) 52 21 - 715 46
Email: info@planatol.de
www.planatol.de

Member of IVK

Company

Year of formation
1932

Size of workforce
120

Ownership structure
Blue Cap AG

Sales channels
Direct sale to industrial customers
Graphical retailers
Representatives abroad worldwide
Subsidiaries abroad

Contact partners
Managing director:
Johann Mühlhauser
Phone +49 (0) 80 31-720-0

Range of Products

Types of adhesives
PVAC, EVA, PU & Acryl Dispersions
EVA, PSA, PO Hotmelts
Reactive PUR Hotmelts
Dextrin, Starch and Latex Adhesives
Special-purpose Adhesives

For applications in the field of
Graphics industry
- Fold-gluing
- Bookbinding
- Print finishing (fugitive adhesives, foil and gloss foil
  adhesives, sealing foils, customised solutions for
  forms, notepad production etc.)
Packaging industry
- Folding box & corrugated cardboard
- End-of-line solutions
- Paper sack production (paper sacks, valve sacks,
  tube sacks, cross-floor sacks, heavy-duty bonding,
  paper and foils)
Wood and furniture industry
- Furniture & kitchens
- Wooden composites (plywood, laminated wood,
  wooden boards, finger jointing, veneer processing)
- Design (door production, window production,
  honeycomb panels, sandwich elements, parquet/
  laminate)
*Special-purpose adhesives for the sandwich
construction industry*
- Construction industry (insulation materials, e.g. for
  roof linings or base plates, interior fittings, e.g. panels, furniture, doors, partitions, flame protection)
- Lightweight construction (caravan, shipbuilding,
  aviation, automotive)
Functionalised and custom-made adhesive solutions
- Construction & home (roof linings, vapour barriers,
  decorative fabrics, mattresses)
- Automotive (combinations of light yet robust materials, decorative fabrics, door trims, seat backrests)
- Textile industry (breathable fabric combinations,
  work/protective clothing, technical textiles,
  functional textiles)
Further products:
Adhesive application systems for job printing, newspaper printing, gravure printing and digital printing
Adhesive application systems for hot and cold-gluing
applications. Copybinder and accessories

POLYMERISATION · SPECIALITY CHEMICALS · SERVICES

POLY-CHEM GmbH
ChemiePark Bitterfeld-Wolfen
OT Greppin, Farbenstraße, Areal B
D-06803 Bitterfeld-Wolfen
Phone +49 (0) 3493 75400
Fax +49 (0) 3493 75404
Email: contact@poly-chem.de
www.poly-chem.de

Member of IVK

Company

Year of formation
2000

Size of workforce
50

Sales channels
Direct sales to industrial partners

Contact partners
Management:
Dr. Jörg Dietrich

Application technology:
Dr. Andreas Berndt

Further information
Toll production, Custom Synthesis

Range of Products

We offer to the coating industry, the chemical industry and as well to the segments paints and varnishes.

The present business activities of POLY-CHEM GmbH are
- Solvent based and solvent-free pressure sensitive acrylic adhesives and acrylic polymers in general
- Synthetic rubber pressure sensitive adhesive
- Production of specialty chemicals
- Contract formulations
- Services such as drum filling and tank leasing
- Trading (export/import)

Raw materials
Crosslinker, softening agents, resins, acrylates

For applications in the field of
- Paper/packaging
- Construction industry, including floors, walls and ceilings
- Textile industry
- Adhesive tapes, labels
- Automotive
- Graphic arts

Polytec PT GmbH
Polymere Technologien
Ettlinger Straße 30
D-76307 Karlsbad
Phone +49 (0) 72 43-604-40 00
Fax +49 (0) 72 43-604-42 00
Email: info@polytec-pt.de
www.polytec-pt.de

Member of IVK

Company

Year of formation
2005 (1967)

Size of workforce
29

Contact partners
Management:
info@polytec-pt.de

Application technology and sales:
info@polytec-pt.de

Further information
Polytec PT GmbH develops, manufactures and distributes special adhesives for applications in electronics, electrical engineering and automotive electronics as well as the solar industry and the manufacture of smart cards. In addition to an extensive portfolio of electrically and thermally conductive adhesives, transparent and UV-hardening products, Polytec PT develops tailored formulations for the most demanding of adhesive applications.

Range of Products

Range of Products
Electrically conductive adhesives
Thermally conductive adhesives
Adhesives for optical assemblies/fiber optics
USP Class VI adhesives for medical devices
High temperature adhesives
Potting compounds
UV-curable adhesives
Surface pretreatment devices

For applications in the field of
Electronics
Mechanical engineering and equipment construction
Automotive industry, aviation Industry

PRHO-CHEM GmbH

Dohlenstraße 8
D-83101 Rohrdorf-Thansau
Phone +49 (0) 80 31-3 54 92-0
Fax +49 (0) 80 31-3 54 92-29
Email: info@prho-chem.de
www.prho-chem.de

Member of IVK

Company

Year of formation
1994

Ownership structure
private property

Contact partners
Management:
Otto Kleinhanß

Application technology and sales:
Otto Kleinhanß

Range of Products

Types of adhesives
Hot melt adhesives
Reactive adhesives
Dispersion adhesives
Glutine glue
Pressure-sensitive adhesives

Types of sealants
PUR sealants
Silicone sealants
MS/SMP sealants

For applications in the field of
Paper/packaging
Graphic design
Automotive industry, aviation industry
Hygiene
Adhesive tapes, labels
Industrial applications

Rampf Polymer Solutions GmbH & Co. KG
Robert-Bosch-Straße 8 – 10
D-72661 Grafenberg
Phone +40 (0) 71 23-9342-0
Fax +49 (0) 71 23-93 42-2444
Email: polymer.solutions@rampf-group.com
www.rampf-group.com

Member of IVK

Company

Year of formation
1980

Size of workforce
100

Managing partners
Dr. Klaus Schamel

Ownership structure
Family owned

Contact partners
Management:
Dr. Klaus Schamel

Application technology and sales:
Felix Neef

Further information
https://www.rampf-group.com/de/
produkte-loesungen/chemical/klebstoffe/

Range of Products

Types of adhesives
1- and 2-component Polyurethane adhesives
2-component Epoxy adhesives
1- and 2-component Silicon adhesives
Thermoplastic and reactive hotmelt adhesives

Types of sealants
PUR sealants
Silicone sealants
MS/SMP sealants

For applications in the field of
Household
Sandwich panels, caravan and truck bodies
Automotive
Filter
Construction
Electronics
Mechanical engineering

RAMSAUER®

Ramsauer GmbH & Co. KG
Sarstein 17
A-4822 Bad Goisern
Phone +43 (0) 61 35-82 05
Fax +43 (0) 61 35-83 23
Email: office@ramsauer.at
www.ramsauer.at

Member of IVK

Company

Year of formation
1875

Head of Business Administration
Klaus Forstinger, Head of Business
Administration

Ownership structure
private

Contact partners
Management:
Andreas Kain, Sales Manager

Further information
Company history:
When Ferdinand Ramsauer purchased a small
chalk quarry near Bad Goisern in 1875, he
already had all the skills typical of successful
people up to our days. He had his mind set
on innovation and was fully focussed on achieving his goals. Within less than 20 years, he
increased the chalk output of his quarry about
100 times and made "Ischler Bergkreide"
("Mountain chalk from Bad Ischl") a well-known
brand name. Ferdinand Ramsauer and his son
Josef - whose name our company bears today -
were genuine marketing pioneers.
Right from the beginning, mountain chalk was
used primarily for the production of putty for
glazing. Initially, the raw material was sold
to putty manufacturers. In 1950, Ramsauer
began manufacturing putty on its own. The
evolution from mining operations to sealant
manufacturer was complete.
With the introduction of thermally insulating
windows, new plastic and elastic sealants
were needed. Ramsauer developed the first
such modified putties as early as in the fif-
ties. Between 1960 and 1976, sealants were
launched under the very well known names of

Range of Products

Types of adhesives
Reactive adhesives
(1 and 2 component solutions)
Solvent-based adhesives
Dispersion adhesives

Types of sealants
Acrylic sealants
Butyl sealants
PUR sealants
Silicone sealants
MS/SMP sealants

For applications in the field of
Wood/furniture industry
Construction industry, including floors,
walls and ceilings
Mechanical engineering and equipment
construction
Automotive industry, aviation industry
Hygiene and clean room application
Household, recreation and office

E9 M, HB, R68, HV72, and HV76. Simultane-
ously, Ramsauer started developing the first
water-soluble products, known as acrylates. In
1972, Ramsauer started manufacturing sealants on a silicone basis. In 1976, production of
PU foam was initiated. A patent for 2-compo-
nent systems was registered in 1998.
Currently, Ramsauer manufactures top-quality
sealants of all systems and modifications, including a new silicone-free sealant on a hybrid
basis and a 2-component silicone system.
When looking back at 135 years of company
history, there have been many changes, fundamental changes. However what has remained
is the broad perspective and focus on innovation that lives on in the present generation in
our company.

Renia Gesellschaft mbH
Ostmerheimer Straße 516
D-51109 Köln
Phone +49 (0) 2 21-63 07 99-0
Fax +49 (0) 2 21-63 07 99-50
Email: info@renia.com
www.renia.com

Member of IVK

Company

Year of formation
1930

Ownership structure
GmbH, family owned

Sales channels
exclusive partners in more than 50
countries

Contact partners
Management:
Dr. Rainer Buchholz

Application technology:
Dr. Julian Grimme

Exportmanager:
Dr. Rainer Buchholz

Range of Products

Types of adhesives
Solvent-based adhesives
Dispersion adhesives

For applications in the field of
Household, recreation and office
Shoe Industry
Health

Rhenocoll-Werk e. K.

Erlenhöhe 20
D-66871 Konken
Phone +49 (0) 63 84-99 38-0
Fax +49 (0) 63 84-99 38-1 12
Email: info@rhenocoll.de
www.rhenocoll.de

Member of IVK

Company

Year of formation
1948

Size of workforce
120

Ownership structure
Joint partnership, Fam. Holding

Subsidiaries
Polska, Czechoslovakia, Belarus, Georgia,
Russia, China, India

Sales channels
Dealer based worldwide

Contact partners
Management:
Werner Zimmermann

Range of Products

Types of adhesives
Hot melt adhesives
Reactive adhesives
Dispersion adhesives
Pressure-sensitive adhesives

Types of sealants
Acrylic sealants
PUR sealants

For applications in the field of
Paper/packaging
Wood/furniture industry
Construction industry, including floors,
walls and ceilings
Household, recreation and office

RUDERER KLEBETECHNIK GMBH

Harthauser Straße 2
D-85604 Zorneding (Munich)
Phone +49 (0) 8106 2421 -0
Fax +49 (0) 8106 2421 -19
Email: info@ruderer.de
www.ruderer.de

Member of IVK

Company

Year of formation
1987

Size of workforce
> 30

Ownership structure
Family company

Sales channels
direct and distribution

Range of Products

Types of adhesives
Hot melt adhesives
Reactive adhesives
Solvent-based adhesives
Dispersion adhesives
Pressure-sensitive adhesives

Types of sealants
Acrylic sealants
PUR sealants
Silicone sealants
MS/SMP sealants

Equipment, plant and components
for surface pretreatment
for adhesive curing
adhesive curing and drying
measuring and testing

For applications in the field of
Wood/furniture industry
Electronics
Mechanical engineering and equipment
construction
Automotive industry, aviation industry
Textile industry
Household, recreation and office

RÜTGERS
Germany GmbH

Varziner Straße 49
D-47138 Duisburg
Phone +49 (0) 2 03-42 96-02
Fax +49 (0) 2 03-42 96-7 62
Email: resins@raincarbon.com
www.novares.de

Member of IVK

Company

Year of formation
1849

Sales channels
own sales force
global agent and distribution network

Contact partners
Director Sales/Marketing
Thomas Reisenauer

Sales Manager
Mai Doan

Range of Products

Raw materials
Hydrocarbon resins:
aromatic
aliphatically mod.
Indene-coumarone
hydrogenated
phenolically modified
pure monomer based

Modifiers:
high boiling solvents

SABA Dinxperlo BV

Industriestraat 3
NL-7091 DC Dinxperlo
Phone +31 315 65 89 99
Fax +31 315 65 32 07
Email: info@saba-adhesives.com
www.saba-adhesives.com

Member of VLK

Company

Year of formation
1933

Size of workforce
170

Managing director
W. de Zwart

Ownership structure
R. J. Baruch, W. F. K. Otten

Subsidiaries
SABA Polska SP. z o. o.
SABA North America LLC
SABA Pacific
SABA China
SABA Bocholt GmbH

Contact:
info@saba-adhesives.com

Further information
Whether you are active in construction or
industry, the adhesives and sealants that you
use have to meet stringent requirements.
SABA is a producer of high-quality and technically progressive adhesives and sealants.
With our knowledge of adhesion and sealing,
we are happy to help you optimise your
processes, so that you produce better end-
products or projects at lower total costs and
in a safer working environment. In this way,
we work together to strengthen your
competitive position.

Read more about SABA:
www.saba-adhesives.com

Range of Products

Types of adhesives
Water-based adhesives
Hot melt adhesives
Solvent-based adhesives
Reactive adhesives
Pressure-sensitive adhesives

Types of sealants
Polysulfide sealants
PUR sealants
Silicone sealants
MS/SMP sealants

Equipment, plant and components
for conveying, mixing, metering and
for adhesive application
for surface pretreatment

For applications in the field of
furniture production
mattress production
foam converting
automotive
pvc bondings
building & construction
marine
transport
civil & environmental engineering

Schill+Seilacher „Struktol" GmbH
Moorfleeter Straße 28
D-22113 Hamburg
Phone +49 (0) 40-733-62-0
Fax +49 (0) 40-733-62-297
Email: polydis@struktol.de
www.struktol.de

Member of IVK

Company

Year of formation
1877

Size of workforce
250 employees in Hamburg

Ownership structure
privately owned

Subsidiaries
Schill+Seilacher, Böblingen (Germany)
Schill+Seilacher Chemie GmbH,
Pirna (Germany)
Struktol Company of America, Ohio (USA)
Struktol Co. Company of America, Georgia (USA)
SNS Nano Fiber Technology LLC., Ohio (USA)

Sales channels
Germany: Direct
International. Distributeurs and Agencies

Contact partners
Dr.-Ing. Hauke Lengsfeld
(General Manager Epoxy Products)
Phone: +49 (0) 40 733 62 268
Email: hlengsfeld@struktol.de

Sven Wiemer
(Senior Manager Epoxy Products)
Phone: +49 (0) 40-733-62-125
Email: swiemer@struktol.de

Further information
Manufacturing and development of tailor-made
solutions of exclusive epoxy prepolymers
and 2C epoxy based composite systems in
cooperation with our customers.

Range of Products

Raw materials
Epoxy Prepolymers
Struktol® Polydis®
Struktol® Polycavit®
Struktol® Polyvertec®
Struktol® Polyphlox®

The product ranges Struktol® Polydis®,
Polycavit® and Polyvertec® are adducts of either
rubber or elastomer modified epoxy resins to
improve the mechanical properties, such as Impact Resistance, T-Peel and Lap Shear next to a
general improvement of the adhesion behavior.

The Struktol® Polyphlox® range consists of
adducts of organo phosphorous modified epoxy
resins to equip epoxy matrices with flame
retardancy.

Applications
Epoxy based
(Structural) Adhesives
Castings
Prepregs
Composites (Hand lay-up, RIM/RTM, SMC/
BMC, Pultrusion)
Fiber-Reinforced Plastics
Flame Ratardancy

Schlüter-Systems KG

Schmölestraße 7
D-58640 Iserlohn/Germany
Phone +49 (0) 23 71-971-0
Fax +49 (0) 23 71-971-111
Email: info@schlueter.de
www.schlueter-systems.com

Member of IVK

Company

Year of formation
1966

Size of workforce
more than 1,600 employees worldwide

Managing partners
Schlüter-Systems KG has system alliances with leading construction chemistry and ceramics companies.

Our constructions chemistry partners are: ARDEX GmbH, PCI Augsburg GmbH, Sopro Bauchemie GmbH, MAPEI GmbH, Kiesel Bauchemie GmbH u. Co. KG and SCHÖNOX GmbH, RAK Ceramics GmbH and V&B Fliesen GmbH.

Subsidiaries
Apart from its head office in Iserlohn, Germany, the company has seven subsidiaries in Europe and North America as well as several service bureaus and distribution partners around the world. In total the company employs more than 1,600 people.

Sales channels
B2B

Contact partners
Management:
Günter Broeks (Sales Director)

Application technology:
Rainer Reichelt
(Head of International Technical Network)

Further information
From the Schlüter-SCHIENE to complete systems – with innovative ideas and high-quality products Schlüter-Systems KG is the market leader for many tile related products.

Range of Products

Types of adhesives
Reactive adhesives
Dispersion adhesives

Types of sealants
Other

Raw materials
Polymers

Equipment, plant and components
for surface pretreatment
for adhesive curing
adhesive curing and drying

For applications in the field of
Construction industry, including floors, walls and ceilings
Adhesive tapes, labels
Hygiene

Schlüter-Systems KG with its headquarter in Iserlohn, Germany, offers complete systems with a portfolio of more than 10,000 products and product types and has distribution partners around the world. Among other things, the family-run company offers drainage systems for balconies and terraces, underfloor heating systems, barrier-free showers, construction panels, uncoupling and insulation solutions and a multitude of profiles. Innovative illuminated profile technology and wall heating systems enlarge Schlüter-Systems KG product portfolio.

With more than 1,600 employees in Europe and the USA Schlüter-Systems KG sets standards in creating solutions for tile setters worldwide.

Schomburg GmbH & Co. KG

Aquafinstraße 2 – 8
D-32760 Detmold
Phone +49 (0) 52 31-9 53-00
Fax +49 (0) 52 31-9 53-1 23
Email: info@schomburg.de
www.schomburg.de

Member of IVK

Company

Year of formation
1937

Size of workforce
220 (Germany), 580 worldwide

Managing partners
Albert Schomburg
Ralph Schomburg
Alexander Weber

Nominal capital
3,619 Mio. €

Ownership structure
Family and Management owned

Subsidiaries
31 worlwide:
Poland, Czech Republic, USA, India, Turkey,
Luxemburg, Switzerland, Russia, Netherlands, Slovakia, Italy, etc.

Sales channels
Distribution partners

Contact partners
Management:
Ralph Schomburg
Alexander Weber

Application technology and sales:
Holger Sass
Michael Hölscher

Range of Products

Types of adhesives
Reactive adhesives
Dispersion adhesives
Cement based adhesives

Types of sealants
Acrylic sealants
Polysulfide sealants
PUR sealants
Silicone sealants
Other

Equipment, plant and components
for conveying, mixing, metering and
for adhesive application

For applications in the field of
Construction industry, including floors,
walls and ceilings
Mechanical engineering and equipment
construction

SCIGRIP Europe Ltd.
Unit 22, Bentall Business Park, Glover Road
NE37 3 JD Washington Tyne & Wear, UK
Phone +44 (0) 191 419 6444
Fax +44 (0) 191 419 6445
Email: info@scigrip-europe.com
www.scigrip.com

Member of IVK

Company

Sales channels
Distribution and direct

Contact partners
Management:
Tim Johnson

Application technology and sales:
Alexander Groeger

Further information
SCIGRIP, a global supplier of smart adhesive solutions, delivers the latest advancements in bonding systems and adhesive chemistries for the most demanding applications.

With a dedicated research and development team, SCIGRIP excel at offering customers unique bonding requirements and help to improve component aesthetics, optimize part design and increase productivity.

SCIGRIP now delivers a broader technology platform than ever before and unique adhesive solutions are available for bonding a range of substrates including metals, thermoplastics and thermoset composites.

SCIGRIP offers powerful, industry-leading technologies in 10:1 and 1:1 methyl methacrylate (MMA), anaerobic, cyanoacrylate (CA) and ultra violet (UV) cure adhesives. All SCIGRIP's formulations are environmentally responsible, low volatile organic compound (VOC) adhesives.

SCIGRIP is committed to manufacturing high quality products and maintains ISO

Range of Products

Types of adhesives
Reactive adhesives

Types of sealants
Silicone sealants
Other

For applications in the field of
Construction industry, including floors, walls and ceilings
Electronics
Mechanical engineering and equipment construction
Automotive industry, aviation industry
Household, recreation and office

9001 certification at all its US and European manufacturing facilities. SCIGRIP has also achieved multiple third party product certifications including Lloyd's Register, GREEN-GUARD, RINA, and BRE.

SCIGRIP Smarter Adhesive Solutions is a wholly owned subsidiary of IPS Corporation, a dedicated, successful adhesives manufacturer for over 65 years. This depth of experience has resulted in an organization that is consistently creating new adhesive chemistries and bonding solutions.

Making a *positive* difference

SCOTT BADER
Wollaston, Wellingborough
Northants, NN29 7 RL
Phone +44 (0) 1933 663100
Email: enquiries@scottbader.com
www.scottbader.com

Company

Founded In
1921

Employees
700

Technical Service Centre
Matt Savage
(matt_savage@scottbader.com)

Sales Manager – Germany, Austria and Switzerland
Manfred Fischer
(manfred_fischer@scottbader.com)

Additional Information
Scott Bader is a global manufacturer and supplier of structural adhesives and polyester/vinylester bonding pastes, with 13 offices and 6 manufacturing sites globally.

Crestabond® MMA adhesives are primer-less, structural adhesives for bonding plastics (including low surface energy thermoplastics), composites and metals. Crestabond® structural adhesives can be used in multiple markets including; automotive, marine, rail, building and construction, wind energy, bus and truck.

Visit www.scottbader.com for more

Range of Products

information.

Types of adhesives
Crestabond® MMA Structural Adhesive

Crestomer® Urethane Acrylate Structural Adhesive

Crestafix® Polyester and Vinylester Bonding Pastes

BUILDING TRUST

Sika Automotive GmbH
Reichsbahnstraße 99
D-22525 Hamburg
Phone: +49 (0) 40-5 40 02-0
Fax: +49 (0) 40-5 40 02-5 15
Email: info.automotive@de.sika.com
www.sikaautomotive.com

Member of IVK

Company

Year of formation
1928

Size of workforce
260

Subsidiaries
Sister companies in 101 countries

Contact partners
Managing Director:
Heinz Gisel

Export:
Kai Paschkowski

Range of Products

Types of adhesives
Hot melt adhesives
Reactive adhesives
Solvent-based adhesives
Dispersion adhesives
Pressure-sensitive adhesives

Types of sealants
PUR sealants
Other

For applications in the field of
Electronics
Automotive industry
Textile industry
Adhesive tapes & labels

**Creating Solutions
for Increased Productivity**
Sika is supplier and development partner to
the automotive industry. Our state-of-the-art
technologies provide solutions for increased
structural performance, added acoustic
comfort and improved production processes. As a specialty company for chemical
products, we concentrate on our core
competencies: **Bonding – Sealing –
Damping – Reinforcing**
As a globally operating company, we are
partner to our customers worldwide. Sika
is represented with its own subsidiaries in
all automobile-producing countries, thus
guaranteeing a professional and fast local
service.

BUILDING TRUST

Sika Deutschland GmbH
Stuttgarter Straße 139
D-72574 Bad Urach
Phone +49 (0) 71 25-940-761
Fax +49 (0) 71 25-940-763
Email: industry@de.sika.com
www.sika.de/industrie

Member of IVK, FKS, VLK

Company

Year of foundation
1910

Size of workforce
20,000 (worldwide Sika AG),
1,500 Sika Deutschland GmbH

Subsidiaries
in more than 100 countries,
see www.sika.com

Sales channels
direct and distribution

Range of Products

Industry and sealants
PUR adhesives
Reactive adhesives
Dispersion adhesives
Solvent-based adhesives
Hot melt adhesives
Pressure-sensitive adhesives
Epoxy adhesives
Acrylic adhesives and sealants
Laminating adhesives
Silicones
SMP adhesives and sealants

Construction
PUR adhesives and sealants
Polysulfide sealants
Acrylic adhesives and sealants
Silicones
SMP adhesives and sealants
Butyl sealants

For applications in the field of
Mechanical engineering and equipment
construction
Construction industry, including floors,
walls and ceilings
Automotive industry, aviation industry
Wood/furniture industry
Electronics
Adhesive tapes, labels
Household, recreation and office
Photovoltaics
Renewable Energies
Marine
Structural glazing
Facades
Building Constructions
Track Construction

BUILDING TRUST

Sika Nederland B.V.
Zonnebaan 56
NL-3542 EG
Phone +31 (0) 30-241 01 20
Fax +31 (0) 30-241 01 20
Email: info@nl.sika.com
www.sika.nl

Member of VLK

Company

Size of workforce
140

Subsidiaries
2

Contact partners
Management:
info@nl.sika.com

Application technology and sales:
info@nl.sika.com

Range of Products

Types of adhesives
Hot melt adhesives
Solvent-based adhesives
Dispersion adhesives

Types of sealants
Acrylic sealants
PUR sealants
Silicone sealants
MS/SMP sealants

Equiment, plant and components
for surface pretreatment

For applications in the field of
Wood/furniture industry
Construction industry, including floors,
walls and ceilings
Electronics
Mechanical engineering and equipment
construction
Automotive industry, aviation industry

SKZ - KFE GmbH

Friedrich-Bergius-Ring 22
D-97076 Würzburg
Phone +49 (0) 931 4104-0
Fax +49 (0) 931 4104-707
Email: kfe@skz.de
www.skz.de

Member of IVK

Company

Year of formation
1961

Size of workforce
181 in 2017

Managing partners
FSKZ e.V.

Nominal capital
35 TEUR

Ownership structure
100% subsidiary

Contact partners
Management:
Mr. Dr. Thomas Hochrein

Application technology and sales:
Mr. Dr. Eduard Kraus

Further information
The German Plastics Center (SKZ) provides independent services in the field of plastics technology for both the plastics industry and public sector. It was founded in Munich in 1961 and today it is one of Europe's largest industry-specific institutes with worldwide monitoring of over 900 plastic products from more than 400 companies.

Range of Products

Types of adhesives
Hot melt adhesives
Reactive adhesives
Solvent-based adhesives
Dispersion adhesives
Vegetable adhesives, dextrin and starch adhesives

Types of sealants
Acrylic sealants
Butyl sealants
PUR sealants
Silicone sealants
MS/SMP sealants

Raw materials
Additives
Fillers
Resins
Solvents
Polymers

Equipment, plant and components
for conveying, mixing, metering and for adhesive application
for surface pretreatment
for adhesive curing
measuring and testing

For applications in the field of
Wood/furniture industry
Construction industry, including floors, walls and ceilings
Electronics
Mechanical engineering and equipment construction
Automotive industry, aviation industry
Adhesive tapes, labels

AUTOMATED SEALING SOLUTIONS

Sonderhoff Chemicals GmbH
Richard-Byrd-Straße 26, D-50829 Cologne
Phone +49 221 95685-0, Fax +49 221 95685-599
Email: info@sonderhoff.com, www.sonderhoff.com

Company

Year of formation
1958

Size of workforce
ca. 100 employees
(Sonderhoff-Group worldwide ca. 330)

Ownership structure
Henkel AG & Co. KGaA

Sister companies
Sonderhoff Engineering GmbH, Dornbirn
(Austria)
Sonderhoff Services GmbH, Cologne
(Germany)
Sonderhoff Polymer-Services Austria GmbH,
Hörbranz (Austria)
Sonderhoff Italia SRL, Oggiono (Italy)
Sonderhoff USA Corporation, Elgin (USA)
Sonderhoff (Suzhou) Sealing Systems
Co.Ltd, Suzhou (China)

Sales channels
worldwide

Contact partners
Application technology and sales:
Daniel Koscielny (sales director)
Dominique Tosi (application engineer)

Further information
Sonderhoff, the system supplier for Formed-
In-Place technology, offers automated
sealing solutions from one hand: more than
1,000 recipes for foam sealing, potting and
gluing, mixing and dosing machines for fully
automated material application as well as

Range of Products

Types of adhesives
2-Component reaction adhesives based on
polyurethane (PU), available in different
degrees of hardness and viscosities (liquid to
stable). Good adhesion on many substrates.
Good temperature and climate change resistance, compliant to automotive requirements,
and also used in others technical sectors.

Types of sealants
2C PU, 2C Silicone, 1C PVC

Systems/Procedures/Accessories/Services
Dispensing systems (for 1-C and 2-C/Multi-
Component systems), linear robots, mixing
and dosing technology for fully automated
adhesive, potting and foam sealing applications as well as services and contract
manufacturing

For applications in the fields
Automotive, packaging, electronics, switch
cabinets, aviation, filter, air condition, lighting,
machine and device construction, appliances,
photovoltaics, solar thermal energy

contract manufacturing for foam sealing,
gluing and potting. OEM's and industrial
suppliers benefit from the great depth of
patented knowledge and many years of
experience from Sonderhoff.

Sopro Bauchemie GmbH
Biebricher Straße 74
D-65203 Wiesbaden
Phone +49 611-1707-239
Fax +49 611-1707-240
Email: international@sopro.com
www.sopro.com

Member of IVK, FCIO

Company

Year of formation
1985 as Dyckerhoff Sopro GmbH, in the year 2002 change in Sopro Bauchemie GmbH

Size of workforce
334 employees

Managing partners
Michael Hecker, Andreas Wilbrand

Subsidiaries
Germany, Hungary, Switzerland, Netherland, Poland, Austria

Sales channels
through specialised distributors

Contact partners
Management:
International Business
Phone +49 611-1707-239
Fax +49 611-1707-240
Email: international@sopro.com

Further information
Sopro offers a comprehensive range of tile-fixing and building chemicals products. Our clear-cut brand strategy has established us as a leading specialist in this sector. Sopro's wide-ranging product portfolio, featuring a consistently high proportion of new products, addresses the full gamut of tiling and building chemicals applications while guaranteeing a product quality that complies, in all respects, with the strict standards set by professional applicators. The builders' merchants sector represents the only acceptable marketing channel for the Sopro brand, Merchants are able to provide both professional tradesmen and the ambitious private client with the standard of counselling appropriate to our high-grade, technologically advanced products.

Range of Products

Types of adhesives
Cementitious adhesives
Reactive adhesives
Dispersion adhesives

Types of sealants
Acrylic sealants
PUR sealants
Silicone sealants

For applications in the field of
Construction industry, including floors, walls and ceilings

Our product range is divided into thress sectors:
Tile fixing products: Tile Adhesives; Tile grouts; Surface fillers; Primers and bonding agents; Waterproofings; Accessories (impact sound insulation and seperating layer systems); Cleaning, impregnation and maintenance; Tiling tools

Building chemical products: Bitumen products; Screeds, binders and construction resins; Mortar and screed additives; Bedding and multi-purpose mortars; Underground construction, gully-repair and shrinkage compounded grouts; Surface fillers; Concrete repairs; Primers, bonding agents, cement paints and silicia sands

Gardening and landscaping products: Drainage and bedding mortars; Paving grouts; Waterproofings; Surface gradient fillers and rapid set mortars; Cleaning, impregnation and maintenance; Landscaping product systems

Stauf Klebstoffwerk GmbH

Oberhausener Str. 1
D-57234 Wilnsdorf
Phone +49 (0) 27 39-3 01-0
Fax +49 (0) 27 39-3 01-2 00
Email: info@stauf.de
www.stauf.de

Member of IVK, FCIO

Company

Year of formation
1828

Size of workforce
81

Ownership stucture
100 % Family Stauf

Managerial head
Volker Stauf, Wolfgang Stauf,
Dr. Frank Gahlmann

Sales channels

Worldwide distribution, own field service and
distributing warehouses in Germany and other
countries for:
• the wood flooring wholesale
• the floor covering wholesale
• the construction material wholesale
• handcraft enterprises
• contractors

Products
• adhesive systems for floor covering and
  wood flooring
• mounting repair adhesives
• artificial turf adhesives
• primers
• levelling compounds
• underlayments
• surface treatment products
• accessories

Contact partners
Management:
Volker Stauf, Phone +49 (0) 27 39-30 10,
info@stauf.de

Product Technology:
Dr. Frank Gahlmann, Phone +49 (0) 27 39-3
01-1 65, frank.gahlmann@stauf.de

Range of Products

Types of adhesives
Polyurethane adhesives
Silane adhesives
Epoxy adhesives
Solvent-based adhesives
Dispersion adhesives

Types of surface treatment products
waterborne coatings for wood flooring
solvent based coatings for wood flooring
Oils
Care + cleaning systems

For applications in the field of
Construction industry, including floors,
walls and ceilings
• Parquet and wood flooring
• End grain wood blocks
• Textile and elastic floor coverings
• Artificial turf
• Mounting repair
• Industrial application

Further information
www.stauf.de

STAUF is a leading system supplier for flooring
technology. For the safe and durable bonding of wood
flooring and floor coverings we research, develop and
produce innovative adhesive systems on a high-grade
raw material basis. Next to the latest adhesive
systems STAUF offers professionel surface treatment
products for wooden surfaces as well as the whole
bandwidth of products for sub floor preparation and
accessories.

STAUF has remained a family-owned company even
after more than 190 years.

Long time experience as well as continuous advance-
ment in a state of the art production and research
environment ensure the constant top-level product
quality and set the standards for the customers of the
wood flooring and floor covering branch.

Stockmeier Urethanes GmbH & Co. KG
Im Hengstfeld 15
D-32657 Lemgo
Phone +49 (0) 52 61-66 0 68-0
Fax +49 (0) 52 61-66 0 68-29
Email: urethanes.ger@stockmeier.com
www.stockmeier-urethanes.com

Member of IVK

Company

Year of formation
1991

Size of workforce
approx. 170

Ownership structure
Member of Stockmeier Group

Subsidiaries
Stockmeier Urethanes USA Inc., Clarks-
burg/USA, Stockmeier Urethanes France
S.A.S.,Cernay, Stockmeier Urethanes UK
Ltd., Sowerby Bridge/UK

Sales channels
direct and distribution

Contact partners
Management:
Stefan Baumann

Application technology and sales:
Frank Steegmanns

Further information
Stockmeier Urethanes is a leading international manufacturer of polyurethane systems.
We are a specialized subsidiary of the
family-owned Stockmeier Group. With four
production plants including R & D in Europe
and the USA we are developing and producing polyurethane systems for sports and

Range of Products

Types of adhesives
Reactive adhesives

Types of sealants
PUR sealants

For applications in the field of
Wood/furniture industry
Construction industry, including floors,
walls and ceilings
Electronics
Mechanical engineering and equipment
construction
Automotive industry, aviation industry

leisure flooring, ACE (Adhesives, Coatings &
Elastomers) and electrical encapsulation
since 1991. In our business unit Adhesives
we are producing systems for manufactures
in the markets Batteries, Filters, Sandwich-
panel, Transportation, Caravan or customized
products for other industrial applications.
Our top brands are Stobielast, Stobicast,
Stobicoll and Stobicoat. More information:
www.stockmeier-urethanes.com

Synthomer Deutschland GmbH
Werrastraße 10
D-45768 Marl
Email: europe.adhesives@synthomer.com
www.synthomer.com

Company

Range of Products

Company

Synthomer is one of the world's leading suppliers of speciality polymers and has leadership positions in many industry segments including adhesives, coatings, construction, technical textiles, paper and synthetic latex gloves. Our polymers help customers to enhance the performance of their existing products and create the next generation products that will fulfil the ever increasingly challenging market needs.

We provide reliable product supply security and customer service worldwide, through a strong network of local commercial and technical service teams, regional R&D and production footprint.

Synthomer has its operational Headquarter in London, UK, and provides customer-focused services from regional centres in Harlow (UK); Marl (Germany); Kuala Lumpur (Malaysia) and Atlanta (USA). It employs more than 2,900 employees across and supply our customers from more than 20 sites.

Contact

Dr. Katja Greiner
European Technical Sales Manager Adhesives
SBU Functional Solutions
Phone: +49 (0) 2365-49 9816
Mobile: +49 (0) 171-813 7422
Email: katja.greiner@synthomer.com

Dr. Graeme Roan
Product Manager Adhesives
SBU Functional Solutions
Mobile: +1 864 398 7260
Email: graeme.roan@synthomer.com

Raw materials

Polymer dispersions and latexes for tapes, labels, protective foils, wood & packaging adhesives and sealants

Pure acrylics *Plextol*®

Styrene acrylics *Revacryl*®

Vinylacetate
homo- and copolymers *Emultex*®

Styrene butadiene latexes *Litex* ®

Redispersible powder *Axilat* ®

Thickeners and other additives *Rohagit* ®

SYNTHOPOL
THE RESIN COMPANY

Synthopol Chemie
Alter Postweg 35
D-21614 Buxtehude
Phone +49 (0) 41 61-70 71 962
Fax +49 (0) 41 61-8 01 30
Email: bprueter@synthopol.com
www.synthopol.com

Member of IVK

Company

Year of formation
1957

Size of workforce
190

Ownership structure
Family company

Sales channels
Germany and Europe

Contact partners
Management:
Dr. Rüdiger Spohnholz
Phone +49 (0) 41 61-70 71 160
Dr. Stephan Reck
Phone: +49 (0) 41 61-70 71 130

Sales:
Hubert Starzonek (Commercial Manager)
Phone +49 (0) 41 61-70 71 770

Application technology:
Dipl. Ing. Rainer Jack
Phone +49 (0) 41 61-70 71 171

Further information
Birgit Prüter
Phone +49 (0) 41 61-70 71 962

Range of Products

Raw materials
acrylic emulsions
polyurethane dispersions
saturated polyesters
solvent based and waterbased acrylics

For applications in the field of
construction adhesives
automobile adhesives
textile adhesives
pressure sensitive adhesives

TER Chemicals Distribution Group
Börsenbrücke 2
D-20457 Hamburg
Phone +49 (0) 40-30 05 01-0
Email: info@terhell.com
www.terchemicals.com

Member of IVK

Company

Year of formation
1908

Size of workforce
421

Managing shareholder
Christian Westphal

Sales revenue
410 Mio. €

Subsidiaries
21

Sales channels
Trading, Distribution, Salesforce

Contact partners
Management:
Andreas Früh, CEO

Application technology and sales:
Jens Vinke

Further information
www.terchemicals.com

Range of Products

Types of adhesives
Hot melt adhesives
Reactive adhesives
Solvent-based adhesives
Dispersion adhesives
Pressure-sensitive adhesives

Types of sealants
Butyl sealants
PUR sealants
PIB

Raw materials
Additives
Fillers
Resins
Solvents
Polymers

For applications in the field of
Paper/packaging
Bookbinding/graphic design
Wood/furniture industry
Construction industry, including floors,
walls and ceilings
Electronics
Automotive industry, aviation industry
Textile industry
Adhesive tapes, labels
Hygiene

tesa SE

Hugo-Kirchberg-Straße 1
D-22848 Norderstedt
Phone +49 (0) 40 88899-0
Fax +49 (0) 40 88899-6060
www.tesa.de
www.tesa.com

Member of IVK

Company

Year of formation
tesa AG 2001,
tesa SE since march, 30th 2009

Size of workforce
4,917

Ownership structure
100 % subsidiary of Beiersdorf AG, Hamburg

Subsidiaries
64

Sales channels
Industry (e. g. automotive, electronics,
print & paper, solar), Food and DIY

Contact
www.tesa.com

Further information
tesa SE is one of the world's leading manufacturers of technical adhesive tapes and
self-adhesive system solutions (more than
7,000 products) for industrial and professional customers as well as end consumers.
Since 2001, tesa SE (4,917 employees) has
been a wholly owned affiliate of Beiersdorf
AG (whose products include NIVEA, Eucerin,
and la prairie). Applications for various industrial sectors, such as the automotive industry,
the electronics sector (e.g. smartphones,
tablets), printing and paper, building supply,
and security concepts for effective brand and
product protection, account for about three-
quarters of the tesa Group's sales (2018:
1.3428 billion euros). tesa also partners with
the pharmaceuticals industry to develop and

Range of Products

Types of adhesives
Hot melt adhesives tapes
Reactive adhesives tapes
Pressure-sensitive adhesives tapes

For applications in the field of
Paper/packaging
Bookbinding/graphic design
Wood/furniture industry
Construction industry, including floors,
walls and ceilings
Electronics
Mechanical engineering and equipment
construction
Automotive industry, aviation industry
Adhesive tapes, labels
Household, recreation and office

---

produce medicated patches. tesa earns just
under one-quarter of its sales with products
for consumers and professional craftsmen,
offering 300 applications for end consumers
that amongst others make working in the
home and the office easier. tesa became
famous for branded products for end consumers - like tesa Powerstrips® or tesafilm®,
one of the few brand names to be listed in
the Duden dictionary. tesa is active in more
than 100 countries worldwide.
Find out more about tesa SE at www.tesa.com

tremco illbruck GmbH

Von-der-Wetter-Straße 27
51149 Köln
Phone: +49 (0) 22 03-57 55-0
Fax +49 (0) 22 03-57 55-90
Email: info.de@tremco-illbruck.com
www.tremco-illbruck.com

Member of VLK

Company

Year of formation
tremco 1928

Size of workforce
> 1,000

Sales channels
• Distributors (Construction business)
• Direct sales to Key Account Customers
  in the manufacturing industry and to EIFS
  and Window & Facade System Providers

Contact partners
Business Unit Industrial Solutions:
Andres Klapper

Further information
tremco illbruck is one of Europe's leading
manufacturers of highperformance sealing
and bonding products. Sealing tapes,
membranes, adhesives and sealants with
particular properties form the core of our
extensive range.
We are specialises in finding individual
solutions for specific requirements from
industrial customers and in optimising their
production processes.

www.tremco-illbruck.com

Range of Products

Types of sealants and adhesives
1-part and 2-part silicones
1-part and 2-part polyurethane
1-part Hybrids
Butyl
Acrylic
Hotmelt

For applications in the field of
Window and Facade Systems
Insulating and Structural Glazing
Exterior Insulation and Finish Systems
Construction, incl. window, facade and
interior construction
Automotive Aftermarket and Transportation
Household appliances
Electronics
Ventilation and air conditioning
Photovoltaic

TSRC (Lux.) Corporation S.a.r.l.

34 – 36 Avenue de la Liberté
L-1930, Luxembourg
Phone + 352-26 29 72-1
Fax + 352-26 29 72-39
Email: info.europe@tsrc-global.com
www.tsrcdexco.com

Member of IVK

Company

Year of formation
2011
(for the Europe branch in Luxembourg)

Size of workforce
14

Contact partners
Management:
Christian Kafka

Application technology and sales:
Beverley Weaver

Range of Products

Types of adhesives
Hot melt adhesives
Solvent-based adhesives
Pressure-sensitive adhesives

Types of sealants
Other (Styrenic Block Copolymers)

Raw materials
Polymers: Styrenic Block Copolymers
(for applications under 1. & 2.)

For applications in the field of
Paper/packaging
Bookbinding/graphic design
Wood/furniture industry
Construction industry, including floors,
walls and ceilings
Automotive industry, aviation industry
Adhesive tapes, labels
Hygiene
Household, recreation and office

Türmerleim AG

Hauptstrasse 15
CH-4102 Binningen
Phone +41 (0) 61 271 21 66
Fax +41 (0) 61 271 21 74
Email: info@tuermerleim.ch
www.tuermerleim.ch

Member of FKS

Company	Range of Products

Year of formation
1992

Size of workforce
8

Managing partners
Marcel Leder-Maeder

Types of adhesives
Hot melt adhesives
Emulsions
Starch, dextrin and casein adhesives
UF-/MUF-resins

For applications in the field of
Paper/packaging
Labelling
Wood/furniture industry
Tissues

Türmerleim GmbH

Arnulfstraße 43
D-67061 Ludwigshafen
Phone +49 (0) 6 21-5 61 07-0
Fax +49 (0) 6 21-5 61 07-12
Email: info@tuermerleim.de
www.tuermerleim.de

Member of IVK, FKS

Company

Year of formation
1889

Size of workforce
130

Managing directors
Matthias Pfeiffer
Dr. Thomas Pfeiffer
Martin Weiland

Subsidiaries
Türmerleim AG, Basel

Contact partners
Management:
Dr. Jörg Liebe
Joset Karl
Tanguy Trippner
Harald Staub

Application technology and sales:
see Management

Range of Products

Types of adhesives
Hot melt adhesives
Emulsions
Starch, dextrin and casein adhesives
UF-/MUF-resins

For applications in the field of
Paper/packaging
Labelling
Wood/furniture industry
Tissues

BOLTON
ADHESIVES

UHU GmbH & Co. KG
Herrmannstraße 7, D-77815 Bühl
Phone +49 (0) 72 23-2 84-0
Fax +49 (0) 72 23-2 84-2 88
E-Mail: info@uhu.de

www.UHU.de
www.UHU-profi.de
www.boltonadhesives.com

Headquarter: Bolton Adhesives
Adriaan Volker Huis – 14th floor
Oostmaaslaan 67
NL-3063 AN Rotterdam

Member of VLK

Company

Year of formation
1905

Size of workforce
Bolton Adhesives > 700 employees

Managing partners
Bolton Adhesives B. V., Bolton Group

Ownership structure
UHU is member of the Bolton Group

Subsidiaries
UHU Austria Ges.m.b.H., Wien (A)
UHU France S.A.R.L., Courbevoie (F)
UHU-BISON Hellas LTD, Pireus (GR)
UHU Ibérica Adesivos, Lda., Lisboa (P)

Contact partners
Managing Director:
Robert Uytdewillegen, Danny Witjes,
Ralf Schniedenharn

Application technology:
Domenico Verrina

Sales:
Stefan Hilbrath

Sales channels
professional trade, hardware stores,
do-it-yourself hypermarkets, modelbuilding
stores, food trade, stationery, department
stores

Range of Products

Types of adhesives
2K-epoxy resin adhesives
Cyanoacrylate adhesives
Hot melt adhesives
Reactive adhesives
Solvent-based adhesives
Dispersion adhesives
Construction adhesives
Sealants

For applications in the field of
Paper/packaging
Wood/furniture industry
Construction industry
Electronics in industry
Automotive industry
Textile industry

UNITECH
Deutschland GmbH

Mündelheimer Weg 51a – 53
D-40472 Düsseldorf
Phone +49 (0) 211 51 62 1987
Email: hendrik.balcke@unitech99.co.kr
www.unitech99.co.kr

Member of IVK

Company

Year of formation
1999

Size of workforce
250

Subsidiaries
South Korea (HQ), Slovakia, Germany,
Turkey, China

Sales channels
Direct/Indirect

Contact partners
Management
Hendrik A. Balcke
Email: hendrik.balcke@unitech99.co.kr

Range of Products

Types of adhesives
Reactive adhesives
Epoxy adhesives (1C and 2C solutions)

Types of sealants
PVC-based sealants
Other

For applications in the field of
Automotive industry
(Bodyshop, Paintshop and Interior Shop
materials)
Ship building
Electronics

Uzin Utz

Uzin Utz AG
Dieselstraße 3
D-89079 Ulm
Phone +49 (0) 7 31-40 97-0
Fax +49 (0) 7 31-40 97-1 10
Email: info@uzin-utz.com
www.uzin-utz.com

Member of IVK, FCIO, FKS

Company

Year of formation
1911

Size of workforce
1,268 (December 2018)

Managing partners
Julian Utz
Philipp Utz
Heinz Leibundgut

Ownership structure
Public company

Subsidiaries
Switzerland, France, Netherland, Belgium,
Great Britain, Poland, Czechia, Austria, USA,
China, Indonesia, New Zealand, Slovenia,
Croatia, Hungary, Serbia, Norway

Sales channels
Trade, Direct marketing

Contact partners
Head of Research and Development
Johanis Tsalos

Head of Sales
Philipp Utz

Further information
Since its foundation in the year 1911, Uzin
Utz AG has developed from a regional
adhesives manufacture to a globally active
full-range system supplier for flooring
systems.

Range of Products

Types of adhesives
Reactive adhesives
Dispersion adhesives
Pressure-sensitive adhesives
Adhesives on PE film carrier

Types of sealants
Acrylic sealants
PUR sealants
Silicone sealants

Equipment, plant and components
for conveying, mixing, metering and
for adhesive application
for surface pretreatment
measuring and testing

For applications in the field of
Construction industry, including floors,
walls and ceilings

Uzin Utz
SCHWEIZ

Uzin Utz Schweiz AG
Ennetbürgerstrasse 47
CH-6374 Buochs
Phone + 41 41 624 48 88
Fax + 41 41 624 48 89
Email: ch@uzin-utz.com
www.uzin-utz.com

Member of FKS

Company

Year of formation
1933

Size of workforce
38 (55 with subsidiary)

Managing partners
Vitus Meier

Ownership structure
Public company
A company of the Uzin Utz Group
since 1998

Subsidiaries
DS Derendinger AG, Thörishaus, Switzerland

Sales channels
Direct sales, wholesale distribution, trade

Contact partners
Management:
Vitus Meier, Managing Director

Application technology and sales:
Hans Gallati, Head of Sales and Marketing
Switzerland

Further information
Uzin Utz Schweiz AG stands for concentra-
ted floor competence. Since its foundation
in the year 1933, the company has developed itself to a leading full-range system
supplier for flooring systems in Switzerland.
With the brands UZIN, WOLFF, Pallmann,
codex, Derendinger and collfox, Uzin Utz
Schweiz AG provides a full comprehensive
range.

Range of Products

Types of adhesives
Reactive adhesives
Solvent-based adhesives
Dispersion adhesives
Adhesives on special film carrier

Equiment, plant and components
for conveying, mixing, metering and for
adhesive application
for surface pretreatment
measuring and testing

For applications in the field of
Construction industry, including floors,
walls and ceilings
Transport (railway, ships)

Versalis S.p.A.

Piazza Boldrini, 1
20097 San Donato Milanese, Italy
Email: info.elastomers.versalis.eni.com
www.versalis.eni.com

Member of IVK through
Versalis International SA -
Zweigniederlassung Deutschland
Düsseldorfer Straße 13
65760 Eschborn, Germany

Company

Year of formation
1957

Size of workforce
5,200

Managing partners
Ing. Marco Chiappani,
Vice President BU Elastomers

Nominal capital
1.364.790.000 Euro

Ownership structure
Eni S.p.A.

Subsidiaries
see website

Sales channels
Versalis sales network,
see website

Contact partners
Management:
Ing. Franco Ossola,
Sales Manager Elastomers + local sales
network in single countries

Further information
See official website:
www.versalis.eni.com

Range of Products

Types of adhesives
Hot melt adhesives
Solvent-based adhesives
Pressure-sensitive adhesives

Raw materials
Polymers

For applications in the field of
Paper/packaging
Bookbinding/graphic design
Wood/furniture industry
Automotive industry, aviation industry
Adhesive tapes, labels
Hygiene
Household, recreation and office

Vinavil S. p. A.

Via Valtellina, 63
I-20159 Milano
Phone +39-02-69 55 41
Fax +39-02-69 55 48 90
Email: vinavil@vinavil.it
www.vinavil.it

Member of IVK

Company

Year of formation
1994

Size of workforce
> 300

Ownership structure
Mapei S. p. A.

Subsidiaries
Vinavil Americas. Corp.
Vinavil Egypt

Sales channels
> 40 representative commercial offices

Contact partners
Management:
Taako Brouwer
Ing. Silvio Pellerani

Application technology and sales:
Dr. Mario De Filippis
Manfred Halbach
Dr. Fabio Chiozza

Further information
Certified acc. to ISO EN 9001, 14001 and
OHSAS 18001

Range of Products

Types of adhesives
Dispersion adhesives
Pressure-sensitive adhesives

Types of sealants
Acrylic sealants

Raw materials
Polymers:
aqueous polymer dispersions,
solid resins and redispersible powders
RAVEMUL®, VINAVIL®, CRILAT®, RAVIFLEX®
and VINAFLEX® based on vinylacetate, vinylacetate copolymers, vinylacetate ethylene
copolymers, acrylic and styrene acrylic

For applications in the field of
Paper/packaging
Bookbinding/graphic design
Wood/furniture industry
Construction industry, including floors,
walls and ceilings
Automotive industry, aviation industry
Textile industry
Adhesive tapes, labels
Household, recreation and office

visions in tapes

VITO IRMEN GmbH & Co. KG
Mittelstraße 74 – 80
D-53424 Remagen
Phone +49 (0) 26 42-4 00 70
Email: info@vito-irmen.de
www.vito-irmen.de

Member of IVK

Company

Year of formation
1907

Size of workforce
85

Managing director
Dr. Michael Büchner

Nominal capital
4,000,000 €

Ownership structure
Limited commercial partnership

Subsidiaries
Representatives in Poland, Austria, Spain, Netherlands, Czech Republic, Hungary, Russia

Sales channels
direct and via dealers & distributors

Contact partners
Management:
Dr. Michael Büchner

Application technology and sales:
Dr. Michael Büchner

Further information
www.vito-irmen.de

Range of Products

Self-adhesives tapes coated with
Hot melt adhesives
Solvent-based adhesives
Dispersion adhesives
Pressure-sensitive adhesives

For applications in the field of
- Adhesive tapes, labels
- Automotive industry/aviation industry
- Construction industry, including floors, walls and ceilings
- Device for production, storage and transport of glass
- Electronic Assembly
- Hygiene
- Mechanical engineering and equipment construction
- Solar Industry
- Structural Glazing
- Wood/furniture industry

Wacker Chemie AG
Hanns-Seidel-Platz 4
D-81737 Munich
Phone +49 (0) 89-62 79-17 41
Fax +49 (0) 89-62 79-17 70
Email: info@wacker.com
www.wacker.com

Member of IVK

Company

Year of foundation
1914

Size of workforce
about 14,500 employees
on December 31, 2018

Ownership structure
Stock corporation ("Aktiengesellschaft")

Subsidiaries
24 production sites

Subsidiaries and sales offices in 32 countries
in Europe, the Americas and Asia.

Range of Products

Raw Materials
Vinyl acetate polymers:
dispersions, dispersible polymer powders
and solid resins (VINNAPAS® and VINNEX®)
Vinyl acetate/ethylene copolymers:
dispersions and dispersible polymer
powders (VINNAPAS® and VINNEX®)
VC copolymers (VINNOL®)
Silicones

Additives:
Pyrogenic silica (HDK®)
Silanes, adhesion promoters and cross-
linkers (GENIOSIL®)
Foam-control agents and silicone
surfactants
Nanoscale silicone particles for modifying
adhesives (GENIOPERL®)

Sealants and Adhesive Grades
RTV-1 Silicones (ELASTOSIL®)
RTV-2 Silicones
LSR Silicones
Silicone gels and silicone foams
UV-curing systems
Hybrid adhesives (GENIOSIL®)

Anspruch verbindet

Wakol GmbH
Bottenbacher Straße 30
D-66954 Pirmasens
Phone +49 (0) 63 31-80 01-0
Email: info@wakol.com
www.wakol.com

Member of IVK, FCIO, FKS

Company

Year of formation
1934

Size of workforce
350

Management
Steffen Acker
Christian Groß (CEO)
Dr. Martin Schäfer

Subsidiaries
Ditzingen/Germany, Austria, Switzerland,
Poland, Italy, USA, China, Brasil

Sales channels
direct distribution, specialised trade

Range of Products

Types of adhesives
Dispersion-based adhesives &
Sealing Compounds
Reactive adhesives
Solvent-based adhesives
Hot melt adhesives
PVC Plastisols
PU Foam

For applications in the field of
Construction industry, including floors,
walls and ceilings
Automotive industry, Aviation industry,
including passenger seats,
Wood/furniture industry, Metal packaging
industry, Footwear & Leather industry

WEICON GmbH & Co. KG
Koenigsberger Straße 255
D-48157 Muenster
Phone +49 (0) 2 51-93 22-0
Fax +49 (0) 2 51-93 22-2 44
E-mail: info@weicon.de
www.weicon.de

Member of IVK

Company

Year of formation
1947

Size of workforce
263

Subsidiaries
WEICON Middle East L.L.C
WEICON Kimya Sanayi Tic. Ltd. Sti.
WEICON Inc.
WEICON Romania SRL
WEICON SA (Pty) Ltd
WEICON South East Asia Pte Ltd
WEICON Czech Republic s.r.o.
WEICON Iberica S.L.
WEICON Italia S.r.l.

Sales channels
Technical Distributors, Industry directly

Contact partners
Management:
Ralph Weidling
Timo Gratilow

Application technology and sales:
Holger Lütfring
Product Manager

Vitali Walter
Sales Manager

Range of Products

Types of adhesives
2-comp. adhesives
Basis: Epoxy resins, PUR, MMA
1-comp. adhesives
Basis: Cyanoacrylate, PUR, MMA, POP
Reactive adhesives
Solvent-based adhesives

Types of sealants
PUR-Sealants
Silicone sealants
MS/SMP-Sealants

For applications in the field of
Paper/packaging
Wood/furniture industry
Construction industry, including floors, walls
and ceilings
Electronics
Mechanical engineering and equipment
construction
Automotive industry, aviation industry
Household, recreation and office
Mechanical and apparatus engineering
Metal and plastics industry
Marine industry
Automotive supply industry
Vehicle construction
Engine and transmission construction
Advertising technique
Electrical, Electronics
Wood/furniture industry
Construction industry, including floors, walls
and ceilings
Construction
Aviation industry
Household, recreation and office

**Weiss Chemie + Technik
GmbH & Co. KG**
Hansastraße 2
D-35708 Haiger
Phone +49 (0) 27 73-8 15-0
Fax +49 (0) 27 73-8 15-2 00
Email: ks@weiss-chemie.de
www.weiss-chemie.de

Member of IVK, GEV

Company

Year of formation
1815

Size of workforce
325 members of staff within the group

Managing partners
WBV – Weiss Beteiligungs- und Verwaltungs-
gesellschaft mbH

Subsidiaries
Haiger, Herzebrock, Niederdreisbach,
Monroe NC (USA)

Nominal capital
2 Mio. €

Ownership structure
Family share holders

Management
Jürgen Grimm

Contact partners
Reception: Phone +49 (0) 27 73-8 15-0
Sales: Phone +49 (0) 27 73-8 15-2 02
Technology: Phone +49 (0) 27 73-8 15-2 55
Purchase: Phone +49 (0) 27 73-8 15-2 41

Sales channels worldwide
Own sales force, specialist distributors
industry and trade, customers with private
label

Range of Products

Business division adhesives

Types of adhesives
Reactive adhesives (PUR, Epoxy)
Cyanoacrylate instant glues
Hybrid adhesives (STP/MS)
Solvent-based adhesives
Dispersion adhesives
Pressure sensitive adhesives
Cleaners, solvent- and tenside based

For applications in the field of
Window- and door industry
(plastic materials, metal, wood)
Airtight building envelope according to EnEV
Dry construction
Transportation/commercial vehicles,
Shipbuilding, railway vehicles,
Caravan industry, container construction
Fire protection
Air-conditioning technology and ventilation
engineering
Sandwich- and composite panels
Building trade
Wood-/furniture industry

Business division sandwich elements
Light sandwich constructions as heat- and
sound insulating sandwich elements applied
in fields like doors, windows, gates, booth
constructions, automotive industry etc.

Willers, Engel & Co. GmbH

Lippeltstraße 1
D-20097 Hamburg
Phone +49 (0) 40 33 7967-0
Fax +49 (0) 40 33 1980
Email: info@willersengel.de
www.drt-france.com

Member of IVK

Company

Year of formation
1910

Size of workforce
8

Nominal capital
400,000 €

Ownership structure
100 % subsidiary of D.R.T.

Contact partners
Management:
Jens Döhle

Application technology and sales:
Lars-Olaf Jessen
Rüdiger Heidorn

Further information
Sales of:
Rosin, Rosin Derivatives, Resin Dispersions
Terpene Phenolic Resins, Polyterpene Resins
Terpenes

Range of Products

Raw materials
Resins
Solvents

For applications in the field of
Paper/packaging
Bookbinding/graphic design
Wood/furniture industry
Construction industry, including floors, walls and ceilings
Electronics
Mechanical enineering and equipment construction
Automotive industry, aviation industry
Textile industry
Adhesive tapes, labels
Hygiene
Household, recreation and office

Wöllner GmbH

Wöllnerstraße 26
D-67065 Ludwigshafen
Phone +49 (0) 621 5402-0
Fax +49 (0) 621 5402-411
Email: info@woellner.de
www.woellner.de

Member of IVK

Company

Year of formation
1896

Size of workforce
approx. 150 employees

Subsidiaries
Wöllner Austria GmbH - Fabriksstraße 4-6,
8111 Gratwein-Straßengel, Austria

Sales channels
Direct sales

Contact partners
Management:
Dr. Barbara März

Application technology and sales:
Jörg Batz – Head of Sales Business Unit
CCC; Dr. Joachim Krakehl – Divisional
Head of Technical Marketing Head of Sales
Business Unit ISD

Further information
Wöllner GmbH is one of Europe's leading
suppliers of soluble silicates, process chemicals and special additives for industrial
applications. Being a family enterprise with
over 120 years of experience, we have
in-depth expertise in applied chemistry, covering the areas of research, development
and production.

We develop innovative solutions which are
specially geared towards the chemical,
construction, paint and paper industries as
well as many other sectors.

Range of Products

Types of adhesives
Reactive adhesives
Vegetable adhesives, dextrin and starch
adhesives

Raw materials
Additives

For applications in the field of
Paper/packaging
Wood/furniture industry
Construction industry, including floors, walls
and ceilings

Worlée-Chemie GmbH
Grusonstraße 22
D-22113 Hamburg
Phone +49 (0) 40-7 33 33-0
Fax +49 (0) 40-7 33 33-11 70
Email: service@worlee.de
www.worlee.com

Member of IVK

Company

Year established
1962
(founding as subsidary of E.H.Worlée & Co.
established 1851 as trading company)

Employees
about 250
(productions sites in Lauenburg and Lubeck,
regional sales offices in Germany and affiliates
abroad)

Share holder
Dr. Albrecht von Eben-Worlée
Reinhold von Eben-Worlée

Properties/Estate
own by the von Eben-Worlée family
Affiliates:
E.H. Worlée & Co. b.V. Kortenhoef (NL)
E. H. Worlée & Co. (UK) Ltd.
Newcastel-under-Lyme (GB)
Worlée Chemie India Private Limited,
Mumbai(India)
Worlée Italia S.R.L. Mailand (I)
Varistor AG, Neuenhof (CH)
Worlée (Shanghai) Trading Co. Ltd.
Shanghai (CN)

Sales Organisation
regional sales organisations in Germany,
affiliates, distributors and agents abroad

Management
Dr. Albrecht von Eben-Worlée
Reinhold von Eben-Worlée
Joachim Freude

Range of Products

Additives
Acrylic resins
Acrylic dispersions
Alkyd resins
Alkyd emulsions
Polyester
Polyester polyols
Maleic resins
Rosin based hard resins, phenol modified
Adhesion promoter
Special Primer
Pigments
XSBR- water based dispersions of carboxylated
styrene-butadiene latex
HS-SBR water based dispersion of
styrene-butadiene copolymer (high solid)
VA – Vinyl acetate dispersions
VAA – Vinyl acetate copolymer dispersions
PVAC – Polyvinyl acetate resins
Styrene acrylic resins
Butadien acrylic resins
Thickeners
Aliphatic isocyanates
Polyisocyanates
Polythiols
Hydro carbon binder resins

WULFF GmbH u. Co. KG

Wersener Straße 3
D-49504 Lotte
Phone +49 (0) 5404-881-0
Fax +49 (0) 5404-881-849
Email: industrie@wulff-gmbh.de
www.wulff-gmbh.de

Member of IVK

Company

Year of formation
1890

Size of workforce
200

Managing partners
Familiy Israel, Mr. Ernst Dieckmann

Sales channels
directly and handling

Contact partners
Management:
Mr. Alexander Israel, Mr. Jan-Steffen Entrup

Application technology and sales:
Mr. Ralf Hummelt, Dr. Michael Erberich,
Mr. Jörg Dronia

Range of Products

Types of adhesives
Dispersion adhesives

Types of sealants
MS/SMP sealants

For applications in the field of
Construction industry, including floors,
walls and ceilings
Textile industry

ZELU CHEMIE GmbH
Robert-Bosch-Straße 8
D-71711 Murr
Phone +49 (0) 71 44-82 57-0
Fax +49 (0) 71 44-82 57-30
Email: info@zelu.de
www.zelu.de

Member of IVK

Company

Year of formation
1889

Size of workforce
50 employees

Contact for Research & Development/ Application Development
Dr Stefan Kissling

Contact for technical sales
Nathalie Uhrich
Mustafa Türken

Sales channels
Direct sales
Commercial agencies in Germany
and abroad
Global sales

Range of Products

Types of adhesives
Water-based dispersion adhesives
Hotmelt adhesives
Pressure-sensitive adhesives
Solvent-based adhesives (SBS, CR, TPU)
Reaction adhesives

For applications in the field of
Automotive interior
Upholstered furniture
Chairs/office chairs
Foam fabricators
Mattress manufacturing
Automotive and industrial filters
Construction industry
Leather goods manufacturing

Other products
Polyurethane systems for flexible-, visco-,
integral-, semi-rigid-, rigid- and sealing
foams; casting systems

Examples of applications
Bonding of pocket spring and foam mattresses; manufacturing of upholstered furniture;
manufacturing of automotive air and industrial filters; polyurethane systems for head
and arm rests in automotive interior, for bus
and train seats, for seat cushions and arm
rest of office chairs, for automotive air filters,
for technical components for noise and
vibration reduction in automotive industry

Equipment and
Plant Manufacturing

Baumer hhs GmbH
Adolf-Dembach-Straße 19
D-47829 Krefeld
Phone +49 (0) 2151 4402-0
Fax +49 (0) 2151 4402-111
Email: info.de@baumerhhs.com
www.baumerhhs.com

Company

Year of formation
1986

Size of workforce
270 employees worldwide

Managing partners
Baumer Holding AG, Frauenfeld, Schweiz

Ownership structure
Baumer Holding AG, Frauenfeld, Schweiz

Subsidiaries
China, France, India, Italy, Spain, USA,
United Kingdom

Sales channels
Via Headquarter, subsidiaries and dealer
network worldwide

Contact partners
Management:
Percy Dengler, Dr. Oliver Vietze

Application technology and sales:
Director Sales and Service
Roberto Melim de Sousa

Head of Engineering and Technology
Thomas Walther

Further information
Baumer hhs is your leading and competent
partner for industrial solutions in the folding
carton, end-of-line, cigarette packaging,
corrugated, print finishing, braille and
pharmaceutical industries.

Range of Products

System/Processes/Accessory parts/ Services
Gluing systems for cold glue and hot melt
application
Central adhesive feed systems
Quality assurance with sensor and
camera systems
Controllers
High-pressure piston pumps,
diaphragm pumps

For applications in the field of
Paper/packaging
Bookbinding/graphic design
Wood/furniture industry
Corrugated/Folding Carton

Identify potential savings

with GlueCalc

Let's stick together

baumerhhs.com

 **BÜHNEN**

Bühnen GmbH & Co. KG
Hinterm Sielhof 25
D-28277 Bremen
Phone +49 (0) 4 21-51 20-0
Fax +49 (0) 4 21-51 20-2 60
Email: info@buehnen.de
www.buehnen.de

Member of IVK

Company

Year of formation
1922

Size of workforce
97

Ownership structure
Private ownership

Subsidiaries
BÜHNEN Polska Sp. z o. o.
BÜHNEN B. V., NL
BÜHNEN, AT

Contact person
Managing Director & Shareholder:
Bert Gausepohl

Managing Director:
Jan-Hendrik Hunke

International Sales/Marketing
Valentino Di Candido

Sales GER, AT, CH, NL, PL, Intl.
Jan-Hendrik Hunke

Range of Products

Distribution channels
Direct Sales, Distributors

Hot Melt Adhesives
The product range includes a
variety of different hot melt adhesives
for almost every application.
Available bases:
EVA, PO, POR, PA, PSA, PUR, Acrylate.
Available shapes:
slugs, sticks, granules, pillows, blocks,
cartridges, barrels, drums, bags.

Application Technology
Hot melt tank applicator systems with
piston pump or gear pump, PUR- and POR-
hot melt tank systems, PUR- and POR-bulk
unloader, hand guns for spray and bead
application, application heads for bead,
slot, spray and dot application and special
application heads for individual customer
requirements, hand-operated glue applicators, PUR- and POR glue applicators, wide
range of application accessories, customer-
oriented application, solutions.

Applications
Automotive, Packaging, Display Manufac-
turing, Electronic Industry, Filter Industry,
Shoe Industry, Foamplastic and Textile
Industry, Case Industry, Construction
Industry, Florists, Wood-Processing and
Furniture Industry, Labelling Industry.

DREI BOND

Drei Bond GmbH
Carl-Zeiss-Ring 13
85737 Ismaning, Germany
Phone +49 (0) 89-962427 0
Fax +49 (0) 89-962427 19
Email: info@dreibond.de
www.dreibond.de

Member of IVK

Company

Year of formation
1979

Number of employees
52

Partners
Drei Bond Holding GmbH

Share capital
€ 50,618

Subsidiaries
Drei Bond Polska sp. z o.o. in Kraków

Distribution channels
Directly to the automotive industry
(OEM / tier 1/tier 2); indirectly
via trading partners as well as select private
label business

Contacts
Management:
Mr. Thomas Brandl

Application engineering, adhesive and
sealants:
Johanna Wiethaler, Christian Eicke

Application engineering, metering technology:
Sebastian Schmidt, Marko Hein

Adhesive and sealant sales:
Thomas Hellstern, Harald Jost, Christian Eicke

Metering technology sales:
Sebastian Schmidt, Marko Hein

Additional information
Drei Bond is certified according to ISO 9001-
2015 and ISO 14001-2015

Range of Products

Types of adhesives/sealants
- Cyanoacrylate adhesives
- Anaerobic adhesives and sealants
- UV-light curing adhesives
- 1C/2C epoxy adhesives
- 2C MMA adhesives
- 1C MS hybrid adhesives and sealants
- 1C synthetic adhesives and sealants
- 1C silicone sealants

Complementary products:
- Activators, primers, cleaners

Equipment, systems and components
- Drei Bond Compact metering systems →
  semi-automatic application of adhesives and
  sealants, greases and oils
  Metering technology: pressure/time and
  volumetric
- Drei Bond Inline metering systems →fully
  automated application of adhesives and
  sealants, greases and oils
  Metering technology: pressure/time and
  volumetric
- Drei Bond metering components:
  Container systems: tanks, cartridges, drum
  pumps
  Metering valves: progressive cavity pumps,
  diaphragm valves, pinch valves, spray
  valves, rotor spray

For applications in the following fields
- Automotive industry/automotive suppliers
- Electronics industry
- Elastomer/plastics/metal processing
- Mechanical and apparatus engineering
- Engine and gear manufacturing
- Enclosure manufacturing (metal and plastic)

HARDO-Maschinenbau GmbH
Grüner Sand 78
32107 Bad Salzuflen
Germany
Phone +49 (0) 5222-9301-5
Fax +49 (0) 5222-93016
Email: thermo@hardo.eu
www.hardo.eu

Company

Year of formation
1935

Size of workforce
70 employees

Ownership structure
Family owned

Nominal capital
1.022.500 €

Sales channels
Direct sales and distribution partners
worldwide

Contact partners
Managing director:
Ind. & Econ. Engineer Ingo Hausdorf

Sales management:
Hauke Michael Immig

Sales and application technology:
Ralf Drexhage
Reinhard Kölling
Carsten Schoeler
Thomas Sonnenberg

Further information
More than 50 years experience in developing
and manufacturing of adhesive application
and premelting solutions.

Range of Products

Application systems for
Hotmelts
Pressure sensitive hotmelts
Reactive hotmelts
Dispersion adhesives
Reactive adhesives
Water based primers
Butyl
Bituminous masses
Waxes
Diverse substances

Premelting systems for
Hotmelts
Reactive hotmelts

Roller laminating systems
Tape presses
Plate presses
Winding systems

Customer-specific solutions
HARDO is your ideal partner when it comes
to customer-specific solutions. Our team
of application technicians and engineers
will work together with you to create the
optimum machine or system for your work
process. Our laboratory will identify the
best possible application system for your
particular case.

Hilger u. Kern GmbH
Metering and Mixing Technology
Käfertaler Straße 253
D-68167 Mannheim
Phone: + 49 621 3705-500
Fax: + 49 621 3705-200
Email: info@dopag.com
www.dopag.com

Company

Founding Year
1927

Workforce
> 350 worldwide

International Sales and Service
DOPAG Dosiertechnik und Pneumatik AG,
Switzerland
DOPAG S.A.R.L., France
DOPAG UK Ltd., Great Britain
DOPAG Italia S.r.l.
DOPAG (US) Ltd.
DOPAG India Pvt. Ltd.
DOPAG (Shanghai) Metering Technology Co.
Ltd., China
DOPAG Eastern Europe s.r.o., Czech
Republic
DOPAG Korea
DOPAG Mexico Metering Technology SA de CV

Services
- In-house technical centre
- Testing of metering and mixing equipment
- Maintenance
- Repairs
- Spare parts
- Consumables

Further business units
Hilger u. Kern Industrial Technology

Range of Products

Metering and Mixing Technology
- Metering, mixing and dispensing systems for the application of single and two-component materials, e.g. lubricants, adhesives, sealants and potting materials
- Gear and piston pump technology
- Automated dispensing systems: standard and line integration-capable production cells, custom-built dosing systems
- Static and static-dynamic mixing systems
- Dynamic mixing head for gasketing, bonding, sealing, and potting applications with PU and silicone

Metering Components and Pumps
- Drum pumps, barrel pumps
- Metering valves, dispensing valves
- Material pressure regulators
- Gear flow meter

Applications
- Greasing and oiling
- Bonding and sealing
- Electronic potting
- Composites
- Injection moulding (LSR)
- Gasketing

**Innotech Marketing und
Konfektion Rot GmbH**
Schönbornstraße 8
D-69242 Rettigheim
Phone +49 (0) 7253-98 88 55 50
Fax +49 (0) 7253-932 40 77
Email: verkauf@innotech-rot.de
www.innotech-rot.de

Member of IVK

Company

Year of formation
1995

Size of workforce
22

Managing partners
Joachim Rapp, Anja Gaber

Nominal capital
100.000 €

Ownership structure
GmbH

Subsidiaries
Adhetek GmbH

Sales channels
technical trade, direct sales, independent sales
representatives, trade shows

Contact partners
Management: Joachim Rapp, Anja Gaber

Technology/Application equipment: Martin Deutsch

Application Solutions/Adhesive Accessories:
Nadine Knörr

Further information
Innotech offers a wide range and expertise in the
field of bonding and sealing. The company is
specialized on applicators and adhesive accessories from all leading manufacturers. Competent
support & advice, trade, training courses, marketing samples, consulting and repair service
completes Innotech's wide range and expertise in
bonding and sealing. New at Innotech: Bonding
trainings as an official cooperation partner of
Fraunhofer IFAM for European Adhesive Bonder
(EAB) and European Adhesive Specialist (EAS).

Range of Products

Applicators
As distributor and service partner of international leading sealant and adhesive manufacturers, Innotech offers more than 600 models
of applicators and dispensing tools – quality
made in Europe and USA- with more than
200.000 spare parts. This involves the repair
and warranty service as well as the distribution of specialized tools, just-in-time delivery,
consumer parts and accessories.

Adhesive accessories
Wide assortment and worldwide dispatch of
mixers and nozzles and surface treatment
from all leading manufacturers.

Services
Support and advice for applicators, Repair service
for applicators, Support, solutions and training,
Cooperation partner with IFAM – Training Center
for European Adhesive Specialist (EAS) and
European Adhesive Bonder (EAB), Conduction of
training sessions of European Adhesion Specialist
(EAS) & European Adhesive Bonder (EAB),
Logistic and transportation, Sample logistics,
Development of special / tailormade nozzles,
Product innovations

Trade
Adhesives, Special applicators, Cartridges
Mixers and nozzles, Adhesive supplies, Cleaners,
Primer equipment, Surface treatment Pyrosil

Production
Marketing bonding, Refilling service
(in different quantities), Test pieces, Packaging/
Assembling

IST METZ GmbH
Lauterstraße 14 – 18
D-72622 Nürtingen
Phone +49 (0) 0 70 22 - 6 00 20
Fax +49 (0) 0 70 22 - 6 00 276
Email: info@ist-uv.com
www.ist-uv.com

Member of IVK

Company

Year of formation
1977

Size of workforce
550 employees worldwide

Ownership structure
GmbH

Subsidiaries
eta plus electronic GmbH
Integration Technology Ltd.
IST France sarl
IST Italia S.r.l.
IST (UK) Limited
IST Nordic AB
IST Benelux B.V.
UV-IST Ibérica SL
IST America Corp.
IST METZ SEA Co., Ltd.
IST METZ UV Equipment China Ltd. Co.
IST East Asia K.K.

Contact partners
Management:
Christian-Marius Metz

Application technology and sales:
Arnd Riekenbrauck

Range of Products

Equipment, plant and components
for adhesive curing
adhesive curing and drying
measuring and testing

For applications in the field of
Paper/packaging
Building industry
Construction industry, including floors,
walls and ceilings
Electronics
Automotive industry, aviation industry
Adhesive tapes, labels

Nordson Deutschland GmbH
Heinrich-Hertz-Straße 42
D-40699 Erkrath
Phone +49 (0) 211 92 05-0
Fax +49 (0) 211 25 46 58
Email: info@de.nordson.com
www.nordson.com

Company

Year of formation
1967

Size of the workforce
450 employees with Nordson Engineering
(Lüneburg)

Shareholders
Nordson Corporation, USA

Contact partners
Board of management:
Ulrich Bender, Georg Gillessen,
Gregory Paul Merk

Sales
General sales manager:
Georg Gillessen
OEM support:
Michael Lazin
Packaging/assembly applications:
Georg Gillessen

Industrial applications: Jörg Klein
Nonwoven: Kai Kröger
Industrial Coating Systems: Ralf Scheuffgen
Automotive/Sealant Equipment: Volker Jagielki

Sales routes
Through field service staff of Nordson
Deutschland GmbH

Further information
Development centres and production facilities
(ISO-certified) in the USA and Europe, over 7,500
employees, subsidiaries on every continent. In
cooperation with the customer, Nordson develops
complete solutions with integrated systems
and matching components which grow with the
customers' demands.

Range of Products

Equipment and systems for the application of
adhesives and sealants and for surface finishing
with paints, lacquers, other liquid materials or
powders. Nordson systems can be integrated into
existing production lines.

Packaging and assembly applications
Complete adhesive application systems (hot
melt/cold adhesive) for integration into packaging lines. In the field of assembly applications,
Nordson optimises production processes in many
different branches of industry.

Industrial applications
Adhesive and sealant applications for a wide
variety of branches of industry, e.g. filter bonding,
sealing of vehicle rear lights and mattress bonding. Wood processing, e.g. profile wrapping, edge
banding and post-forming.

Nonwoven
Nonwoven (tailored systems for the application
of adhesive and super-absorbent powder for the
production of babies' nappies, slip inlays, sanitary
towels and incontinence articles).

Powder & Liquid Coating
Applications and systems for coating with paints,
lacquers, other liquid materials and powders.

Electronics
Automatic coating and metering plants for the
electronics industry for precise application of
adhesives, grouting compounds, solder pastes,
fluxes, protective lacquers, etc.

Automotive
Engineered systems for the application of structural adhesives and sealants in harsh automotive
manufacturing environments.

Battery Manufacturing
Dispensing systems for 1-part and 2-part
materials used in storage or vehicle battery cell
manufacturing.

 **plasmatreat**

Plasmatreat GmbH
Queller Straße 76 – 80
D-33803 Steinhagen
Phone +49 (0)5204-9960-0
Fax +49 (0)5204-9960-33
Email: mail@plamatreat.de
www.plasmatreat.de

Member of IVK

Company

Founded in
1995

Number of employees
Approx. 250 (worldwide)

CEO
Christian Buske

Distribution channels
Direct sales, subsidiaries, sales partners

Contact
Executive management:
Christian Buske

Sales director Germany:
Joachim Schüßler

Further information
Atmospheric pressure plasma is a key technology for the environmentally friendly and highly
effective pretreatment and functional coating of
material surfaces. Through the development
of a special jet technology, in 1995 Plasmatreat
became the first company in the world to integrate plasma into series production processes
'inline' under normal pressure, thus making it
suitable for use on an industrial scale. The
patented Openair-Plasma® technology is now
used worldwide in virtually all sectors of industry
and has acquired an estimated market share of
approximately 90 percent. As the international
market leader, Plasmatreat invests around
twelve percent of its annual turnover in research
and development. The company collaborates
with the German Federal Ministry of Education
and Research (BMBF) on numerous research

Range of Products

**Systems/processes/accessories/
services**
Surface pretreatment: Microfine cleaning
and simultaneous activation, functional
coatings

Areas of application
Packaging
Furniture
Glass processing, windows
Electronics
Machinery and appliance manufacturing
Automotive engineering
Truck construction
Aerospace industry
Shipbuilding
Medical engineering
New energies (solar technology, wind
power, electromobility)
Consumer goods
Textiles
Adhesive tapes, labels
Hygiene applications
Household appliances, white goods

projects and also works closely with the
Frauenhof Institutes and other leading research
institutes and universities around the world.

The Plasmatreat Group has technology
centers in Germany (head office), the USA,
Canada, China and Japan as well as
subsidiaries and sales partners in 35
countries.

Reinhardt-Technik GmbH
a Member of WAGNER GROUP
Waldheimstraße 3
D-58566 Kierspe
Phone +49 (0) 2359 666-0
Fax +49 (0) 2359 666-129
Email: info-rt@wagner-group.com
www.reinhardt-technik.com

Company

Year of formation
1962

Size of workforce
About 100 employees

Sales Director
Christian Hose (Commercial Director)
Axel Huwald (Manager Sales EMEA)

Direct Sales
Sales representatives, agents and distributors

Company Profile
Reinhardt-Technik GmbH specializes in the areas of bonding, sealing and casting including injection molding. The company offers a comprehensive range of machines for metering and mixing technology, which process cold or heated 1K materials and multi-component liquid plastics such as polyurethanes, polysulphides, epoxies, silicones and LSR (Liquid Silicone Rubber). All common process technologies are offered - from pneumatically or hydraulically driven piston pumps, gear metering systems to electrically controlled shot meter systems. In addition, Reinhardt-Technik is a solution provider and supplies complete custom-engineered systems.

Range of Products

Metering and Potting Systems
Highly precise and reliable dosing systems for a variety of applications:
- Shot meter systems
- Gear metering systems

Automated Production Cells together with partners
Convenient, simple as well as safe operation of the dosing and mixing systems along with integration with upstream and downstream processes:
- Standardized robot cells
- Individual and application-specific manufacturing cells

Material Preparation and Feeding
Modular as well as configurable processing and conveying systems:
- Agitators for fast-sedimenting materials
- 1K- or 2K-compents
- Optional booster addition
- 20 and 200 litre feeding units with robust working cylinder in combination with powerful chop check pumps for low viscosity materials
- Optional pressure containers
- Reliable material discharge with low residual material

Mixing Systems
Low-maintenance and powerful mixing systems for individual processing:
- Static mixing systems
- Dynamic mixing systems
- Incl. snuff back valve

Customer Service
- Technical application centre
- Training
- Hotline
- Remote maintenance
- Online shop (spare and wearing parts)
- High spare part availability
- Service worldwide
- Equipment modernization and modification
- Process optimization

ROBATECH
GLUING SOLUTIONS

Robatech AG
Pilatusring 10
CH-5630 Muri AG
Phone (+41) 56 675 77-00
Fax (+41) 56 675 77-01
Email: info@robatech.ch
www.robatech.ch

Member of IVK

Company

Year of formation
1975

Size of workforce
More than 650 employees all over the world

Shareholder
Robatech AG, CH-5630 Muri, Switzerland

Ownership structure
Robatech AG, CH-5630 Muri, Switzerland

Subsidiaries and agencies
Represented in more than 80 countries
worldwide
Germany: Robatech GmbH,
Im Gründchen 2, D-65520 Bad Camberg
Phone +49 (0) 64 34-94 11-0
Fax +49 (0) 64 34-94 11-22

Channel of distribution
Via Head office, subsidiaries and agencies

Contact partners
Management:
Robatech AG, Switzerland:
Martin Meier
Robatech GmbH, Germany:
Eberhard Schlicht, Andreas Schmidt

Application technology and sales:
Robatech AG, Switzerland:
Kishor Butani, Sales Director
Kevin Ahlers, Marketing Director
Robatech GmbH, Germany:
Eberhard Schlicht, Managing Director

Further information
Production facility in Germany, Hong Kong
and Switzerland

Range of Products

Product and sales program of the company
- Glue application system with piston pumps and gear pumps for hotmelt and dispersions, inclusive necessary equipment.
- Small hotmelt application systems up to 5 liter tank capacity
- Medium hotmelt application systems from 5 to 30 liter tank capacity
- Big hotmelt application systems from 55 to 160 liter tank capacity
- Hotmelt application systems for PUR-hotmelts from 3 to 30 liter tank capacity
- Drumunloaders from 50 to 200 liter tank capacity
- Application technics: Bead application, Surface coating, Spray application, spiral application
- Electrical timing and electrical metering
- Cold glue application systems: Pressure tanks and pump systems

Robatech offers a solution for several industries
Packaging Industry
Converting Industry
Graphic Industry
Woodworking Industry
Building Supplies Industry
Automotive Industry
Various Industries

Rocholl GmbH
Schwarzacherstraße 1
D-74858 Aglasterhausen
Phone +49 (0) 6262 91678-0
Fax 49 (0) 6262 91678-10
Email: post@rocholl.eu
www.rocholl.eu

Member of IVK

Company

Year of formation
1977

Ownership structure
Owner/managing director

Sales channels
direct

Contact partners
Management:
Dr. Matthias Rocholl

Application technology and sales:
Dr. Matthias Rocholl

Range of Products

Equipment, plant and components
measuring and testing

Scheugenpflug
Advanced Dispensing Technology

Scheugenpflug AG
Gewerbepark 23
D-93333 Neustadt
Phone +49 (0) 94 45-95 64-0
Fax +49 (0) 94 45-95 64-40
Email: vertrieb.de@scheugenpflug.de
www.scheugenpflug.de

Company

Year of formation
1990

Size of workforce
More than 600 (worldwide)

Managing Board
Christian Ostermeier, Sergej Erbes, Jürgen Wilde

Contact
Email: vertrieb.de@scheugenpflug.de
www.scheugenpflug.de

Subsidiaries
Scheugenpflug Resin Metering Technologies Co.,
Ltd., China
Email: info@scheugenpflug.com.cn
Scheugenpflug Inc., USA
Email: sales.usa@scheugenpflug-usa.com
Scheugenpflug México, S. de R. L. de C. V.
Email: sales.mx@scheugenpflug-usa.com
Scheugenpflug S.R.L., Romania
Email: info@ro.scheugenpflug.de

Sales Partners Worldwide
See Webpage > Company > Locations & Sales
Partners

Company Profile
Technology that sets standards: With 30 years of
experience, Scheugenpflug is one of the leading manufacturers of innovative dispensing technology. With
an additional core competency in process automation,
the product and technology range extends from
cutting-edge material preparation and feeding units
and high performance dispensing systems to custom-
tailored, modular production lines for a wide range of
applications. Scheugenpflug systems are used in the
automotive and electronics industries as well as the
telecommunications sector, medical technology and
energy supply. The company has additional locations
in China, the USA, Mexico and Romania as well as
numerous service locations and sales partners all over
the world.

Range of Products

Dispensers
- Volumetric piston dispensers
- Alternating volumetric piston dispensers
- A volumetric piston dispenser specifically for
  thermally conductive materials
- Gear pump dispensers

Material Preparation and Feeding
- Feeding from cartridges
- Feeding from hobbocks
- Systems for self-leveling potting media

Systems for Atmospheric Dispensing
- Manual work stations with stand or for
  integration
- Dispensing cells
- ProcessModules for integration

Vacuum Potting Systems
- Entry-level systems
- Large vacuum chambers

Customized Systems and Solutions
Customized automation solutions for various adhesive bonding, sealing and potting tasks, based
on the Scheugenpflug modular system (scalable
design)

Services
- Technology Center/Dispensing tests
- Academy
- Technical Services
- After Sales Services/Spare parts
- Rental equipment

AUTOMATED SEALING SOLUTIONS

Sonderhoff Engineering GmbH
Dr. Walter Zumtobel Straße 15, A-6850 Dornbirn
Phone: +43 5572 39 88 10, Fax: +43 5572 39 88 10-55
Email: info@sonderhoff.com, www.sonderhoff.com

Company

Year of formation
1988

Size of workforce
ca. 100 employees
(Sonderhoff-Group worldwide ca. 330)

Ownership structure
Henkel Central Eastern Europe GmbH

Sister companies
Sonderhoff Chemicals GmbH, Cologne
(Germany)
Sonderhoff Services GmbH, Cologne
(Germany)
Sonderhoff Polymer-Services Austria GmbH,
Hörbranz (Austria)
Sonderhoff Italia SRL, Oggiono (Italy)
Sonderhoff USA Corporation, Elgin (USA)
Sonderhoff (Suzhou) Sealing Systems
Co.Ltd, Suzhou (China)

Sales channels
worldwide

Contact partners
Jürgen Thielert (sales director)

Further information
Development, manufacturing and sales of
mixing and dosing systems as well as
automation concepts for semi and fully
automated systems or stand-alone solutions
for the processing and dosing of 1-compo-
nent and 2-component polymer sealing,
potting and adhesive systems.

Range of Products

Types of adhesives
2-Component reaction adhesives based on PU

Types of sealants
2C PU, 2C Silicone, 1C PVC

Systems/Procedures/Accessories/Services
Dispensing systems (for 1-C and 2-C/
Multi-Component systems), linear robots,
mixing and dosing technology for fully
automated adhesive, potting and foam
sealing applications as well as services and
contract manufacturing.

For applications in the field of
Automotive, packaging, electronics, switch
cabinets, aviation, filter, air condition,
lighting, machine and device construction,
appliances, photovoltaics, solar thermal
energy.

You only glue once.

Sonderhoff – the one-stop-shop for material, machine and contract manufacturing

The gluing of your parts must work right away. Our dosing systems with dynamic mixing ensure high process reliability, an exactly maintained mixing ratio of the adhesive components and a precise adhesive application due to a high repeating accuracy of the mixing head positioning.

Benefit from the great depth of our patented knowledge and many years of experience.

www.sonderhoff.com

AUTOMATED SEALING SOLUTIONS

Sulzer Mixpac Ltd.
Ruetistraße 7
CH-9469 Haag
Phone +41 (0) 81 414 70 00
Email: mixpac@sulzer.com
www.sulzer.com

Company

Year of formation
Sulzer Mixpac is a merger of successful, former independent companies with already existing business connections and coopera- tions. The formal acquisition was realized 2007.

Size of workforce
1,000 employees worldwide

Distribution
Direct distribution to producers of adhesives. toll filler and distribution partners for official trade

Contact partners
Switzerland:
Sulzer Mixpac Ltd.
Email: mixpac@sulzer.com

China:
Sulzer Mixpac China
Email: mixpac@sulzer.com

Untited Kingdom:
Sulzer Mixpac UK
Email: mixpac@sulzer.com

United States:
Sulzer Mixpac USA Inc.
Email: mixpac@sulzer.com

Range of Products

Manufacturer and supplier of metering, mixing, coatings and dispensing systems for reactive multi component material, offering comprehensive systems for various cartridge based 2-K applications.

Cartridge based 2-K systems with different volumes between 2.5 ml and 1.500 ml, with mixing rations of 1 : 1, 2 : 1, 3 : 1, 4 : 1 and 10 : 1
Manual, pneumatic and electric dispensers for 1K and 2K adhesives and sealant applications.
Full range of mixers for cartridge and dispensing machine applications
Brands: MIXPAC™, MK™, COX™

ViscoTec Pumpen- u. Dosiertechnik GmbH
Amperstraße 13
84513 Töging a. Inn
Phone +49 (0) 8631 9274-0
Email: mail@viscotec.de
www.viscotec.de

Company

Year of formation
1997

Size of workforce
250 employees (worldwide)

Managing partners
Dipl.-Ing. Georg Senftl,
Dipl.-Ing. Martin Stadler

Subsidiaries
- Töging am Inn, Germany (Headquarter)
- Georgia, USA
- Singapore
- Shanghai, China
- Pune, India

Sales channels
worldwide

Further information
ViscoTec Pumpen- u. Dosiertechnik GmbH primarily deals with systems required for conveying, dosing, applying, filling and emptying fluids ranging from medium to high viscosity.

Range of Products

Equipment, plant and components
Systems for conveying, dosing, applying, filling and emptying fluids ranging from medium to high viscosity:
- Dispenser and dosing systems (applying and administering, 1K-/2K-dosing, potting, filling, process dosing, spraying)
- Emptying and supplying systems
- Preparing systems
- Complete systems
- Dedicated equipment

For applications in the field of
- Automotive
- Aerospace
- Electronics
- General industry
- Renewable energies
- Plastics
- Food
- Cosmetics
- Pharmaceuticals
- Medical technology
- Biotechnology
- 3D printing
- E-mobility

Walther Spritz- und Lackiersysteme GmbH
Kärntner Straße 18 – 30
D-42327 Wuppertal
Phone +49 (0) 2 02-7870
Fax +49 (0) 2 02-7 87-22 17
Email: info@walther-pilot.de
www.walther-pilot.de

Company

Size of workforce
150 employees

Locations
Wuppertal-Vohwinkel
Neunkirchen-Struthütten

CEO
Wilhelm W. Schmidts
Oliver Witt

Sales Manager
René Brettmann
(Germany, Austria, Benelux)

Applications technology
Gerald Pöplau

Distribution channels
Field sales staff in Germany and
representatives in Europe and overseas.

Range of Products

Systems and components for the application
of adhesives, sealants and paints. Functioning
as a systems supplier, WALTHER engineers
customized, all-round solutions. They guarantee the best results over the long run in
terms of economy, user friendliness and
environment protection.

Application
Adhesive spray guns and automated
applicators
Extrusion guns
Metering valves
Spray guns for dots and lines
Multi-component metering and
mixing systems

Material conveyance
Pressure tanks
Diaphragm pumps
Piston pumps
Pump systems for high-viscosity media
Supply stations
Systems for shear-sensitive materials

Overspray exhaust
Spray booths
Filter technology
Ventilation systems

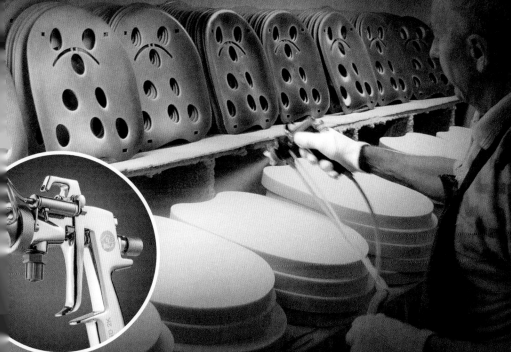

COMPANY PROFILES

Adhesive Technology
Consultancy Companies

ChemQuest Europe Inc.

Bilker Straße 27
D-40213 Düsseldorf
Phone +49 (0) 211-4 36 93 79
Fax +49 (0) 32 12-1 07 16 75
Mobile: +49 (0)1 71-3 41 38 38
Email: jwegner@chemquest.com
www.chemquest.com

Member of IVK

Company

Contact partners
Dr. Jürgen Wegner,
Managing Director
Email: jwegner@chemquest.com

Dr. Hubertus von Voithenberg,
Managing Director
Email: hvoithenberg@chemquest.com

CV and professional background
under www.chemquest.com

Consultancy

The ChemQuest Group is an international
consulting firm headquartering in Cincinnati,
Ohio, USA, with branch offices in Europe,
China, North Africa and Latin America.

We specialize in consulting the Adhesives,
Sealants, Construction Chemicals and Coatings Industry through all steps within the
value chain from raw materials manufacturing through product formulation to all
types of industrial and non-industrial end
use applications. In South Boston/VA we
recently acquired the ChemQuest Technology Institute, a fully equipped testing
and formulating facility for the time being
primarily focusing on coatings, supported by
well experienced professional experts.

Based on in-depth knowledge our service
portfolio includes all types of Management
Consulting, M&A activities, market research, on coatings all types of quality and
regulatory conformity tests, plus market and
technology trend analysis. All associates of
ChemQuest are experienced professionals
from within the Adhesives, Sealants and
Coatings Industry, and we are partnering
with IFAM Fraunhofer Bremen in certified
Adhesives education in North America.
These courses currently take place at our
South Boston facility. For further information and contact details please visit our
website www.chemquest.com.

Our know-how
– your future!

HINTERWALDNER CONSULTING

Consulting Chemists &
Business Economists since 1956

Hinterwaldner Consulting
Dipl.-Kfm. Stephan Hinterwaldner
Markplatz 9
D-85614 Kirchseeon
Phone +49 (0) 80 91-53 99-0
Fax +49 (0) 80 91-53 99-20
Email: info@HiwaConsul.de
www.HiwaConsul.de

Member of IVK

Company

Year of formation
1956

Ownership structure
Privately held

Contact partner
Dipl.-Kfm. Stephan Hinterwaldner
Brigitte Schmid
Andrea Hinterwaldner

Adhesive Consulting
info@HiwaConsul.de

Adhesive Conferencing
contact@mkvs.de
contact@in-adhesives.com

Organizer/Co-organizer

- Munich Adhesives and Finishing Symposium
  Adhesives | Printing | Converting
  www.mkvs.de
- in-adhesives Symposium
  Adhesives | Industrial Adhesives Technology
  Automotive, Electronics, Medicine,
  Lightweight, Composites
  www.in-adhesives.com
- European Coatings Congress

Consultancy

Global Expert Research and Technology
Consulting and Conferencing in the World of
Adhesives

- Raw Materials, Ingredients,
  Intermediates, Additives
- Formulations, Applications,
  Product and Process Developments,
  Feasibility Studies
- Adhesives, Adhesive Tapes, Coatings,
  Cementations, Glues, Sealants,
  Release Liners
- Pressure Sensitive Adhesives,
  Hot Melt Adhesives, Chemically and
  Radiation Curing Adhesives Systems,
  Structural Bonding,
  Structural Glazing Systems
- Technologies in Adhesive Bonding,
  Coating, Converting, Film, Foil, Labeling,
  Laminating, Lightweight, Metalizing,
  Packaging, Printing, Sealing, Surface
- Polymer, Chemically and Radiation
  Curing, Petro-based, Bio-based and
  Green Chemistries
- Cosmetics and Toiletries, Beauty and
  Personal Care, Home Care
- Natural, Renewable, Sustainable,
  Bio-based and Certified Organic Products

Klebtechnik Dr. Hartwig Lohse e. K.

Hofberg 4
D-25597 Breitenberg
Phone +49 (0) 48 22-9 51 80
Fax +49 (0) 48 22-9 51 81
Email: hl@hdyg.de
www.how-do-you-glue.de

Member of the IVK

Company

Year of formation
2009

Contact partners
Dr. Hartwig Lohse

Further information
Based on 30+ years of industrial experience in the development, technical service and technical marketing of adhesives we are providing competent consultancy services along the adhesive supply chain. Our customers are industrial users of adhesives, adhesives manufacturing and raw material supply companies, equipment and plant manufacturing companies as well as manufacturers of dispensing and testing equipment.

Consultancy

Our consultancy service portfolio includes:

- optimisation of current and the planning of new bonding processes incl. generation of the specific requirements profile, selection of the best adhesive technology, source search for commercial adhesives, evaluation of adhesives, bond design advice, selection of suitable surface pre-treatment measures, Development and implementation of the bonding process into production (incl. adhesive dispense, surface pre-treatment, cure process, QC-measures, P-FMEA, etc.).
- implementation of DIN 2304 (Adhesive bonding technology - Quality requirements for adhesive bonding processes)
- adhesive failure analysis (troubleshooting)
- adhesive education, training at all levels specific to the clients requirements
- development of adhesives as well as raw materials for the use in manufacturing of adhesives
- market research, market & market trend analysis
- ...

For further information please visit us at www-how-do-you-glue.de

Contract Manufacturing
and Filling Services

GMBH

LOHNFERTIGUNG UND OPTIMIERUNG

chemisch technische Produkte

Am Nordturm 5
D-46562 Voerde
Phone +49 (0) 2 81-8 31 35
Fax +49 (0) 2 81-8 31 37
Email: mail@loop-gmbh.de

Member of IVK

Company

Year of formation
1993

Size of workforce
25

Contact partners
Managing Director:
DI Marc Zick

Production Manager:
Jürgen Stockmann

Further information
LOOP is consistently expanding its personnel and equipment at plant and the KST Customer Service Test Centre.

LOOP is partner of international and national well known companies.

Range of Products

Toll-Production of the following product groups
- polymer formulations, one- and two- component systems (unfilled, filled), aqueous, solvent based and solvent-free
- additives/concentrates of actives substances (liquid, pasty, powdery)
- impregnating and mould resin systems
- slurries
- powder mixtures
- substrates like SiO_2 coated with active substances
- Concentrate pastes
- Granulates and their fractions for applications in engineering and decorative applications
- compounds for electrical components and diverse additional product groups

LOOP operates on the following sectors for its partners
- adhesives and sealants
- polymer and binder chemicals
- pigments and fillers
- additives
- construction chemicals/preservation of buildings
- composites
- foundry products
- electronics
- cables
- power generation (wind power, photovoltaics)
- glass
- textile and paper finishes
- wood preservatives
- varnishes, paints, and printing inks
- professional development, laboratory, and consultancy services
- large scale distribution, etc.

COMPANY PROFILES

Research and
Development

 **Fraunhofer**

IFAM

Fraunhofer Institute for Manufacturing
Technology and Advanced Materials IFAM
– Adhesive Bonding Technology
and Surfaces –

Wiener Straße 12
D-28359 Bremen
Phone +49 (0) 421 2246-400
Fax +49 (0) 421 2246-430
Email: info@ifam.fraunhofer.de
www.ifam.fraunhofer.de

Member of IVK

Company

Year of formation
1968

Size of workforce
All in all 680
Division of Adhesive Bonding Technology
and Surfaces: > 400

Ownership structure
Fraunhofer IFAM is a constituent entity of
the Fraunhofer-Gesellschaft zur Förderung
der angewandten Forschung e. V. and as
such has no separate legal status.

Contact
Fraunhofer IFAM
**– Adhesive Bonding Technology and
Surfaces –**
Director: Prof. Dr. Bernd Mayer
Deputy director:
Prof. Dr. Andreas Hartwig

Work Areas

R&D – Contract-research and development
– in all fields of adhesive bonding technology and surfaces as well as materials, in addition providing certifying training courses
in adhesive bonding technology and fiber
composite materials:

Adhesives and Polymer Chemistry:
Prof. Dr. Andreas Hartwig
Phone +49 (0) 421 2246-470
Email: andreas.hartwig@ifam.fraunhofer.de

Adhesive Bonding Technology:
Dr. Holger Fricke
Phone +49 (0) 421 2246-637
Email: holger.fricke@ifam.fraunhofer.de

Work Areas

Polymeric Materials and Mechanical Engineering:
Dr. Katharina Koschek
Phone +49 (0) 421 2246-698
Email: katharina.koschek@ifam.fraunhofer.de

*Automation and Production Technology
(CFK Nord, Stade):*
Dr. Dirk Niermann
Phone +49 (0) 4141 78707-101
Email: dirk.niermann@ifam.fraunhofer.de

Adhesion and Interface Research:
Dr. Stefan Dieckhoff
Phone +49 (0) 421 2246-469
Email: stefan.dieckhoff@ifam.fraunhofer.de

Plasma Technology and Surfaces PLATO:
Dr. Ralph Wilken
Phone +49 (0) 421 2246-448
Email: ralph.wilken@ifam.fraunhofer.de

Quality Assurance and Cyber-Physical Systems
Dipl.-Phys. Kai Brune
Phone +49 (0) 421 2246-459
Email: kai.brune@ifam.fraunhofer.de

Paint/Lacquer Technology:
Dr. Volkmar Stenzel
Phone +49 (0) 421 2246-407
Email: volkmar.stenzel@ifam.fraunhofer.de

*Training Center for Adhesive Bonding
Technology:* Dr. Erik Meiß
Phone +49 (0) 421 2246-632
Email: erik.meiss@ifam.fraunhofer.de
www.bremen-bonding.com

*Training Center for Fiber Composite
Technology:* Beate Brede
Phone +49 (0) 421 5665-465
Email: beate.brede@ifam.fraunhofer.de
www.bremen-composites.com

Berner
Fachhochschule

**Institute for Materials and
Wood Technology**
Solothurnstrasse 102, CH-2504 Biel
Phone +41 (0) 32 344 02 02
Fax +41 (0) 32 344 03 91
Email: iwh@bfh.ch
www.bfh.ch/de/forschung/forschungsbereiche/
institut-werkstoffe-holzechnologie-iwh/

Member of IVK

Company

Year of formation
1997

Size of workforce
2,424 employees

Managing partners
Bern University of Applied Sciences

Ownership structure
Governing Body: Canton of Bern

Contact partners
Frederic Pichelin
Email: frederic.pichelin@bfh.ch

Application technology and sales
Martin Lehmann
Email: martin.lehmann@bfh.ch

Further information
Focusing on the sustainable use of resources,
at the Institute for Materials and Wood
Technology of the Bern University of Applied
Sciences, we develop and optimise multi-
functional wood and composite materials
and innovative products for the timber and
construction industries. These products
and materials are sourced from bio-based
materials such as wood and other renewable raw materials. We use our extensive
knowledge of these raw materials to help
us find innovative new uses for them.
In our work, we place particular emphasis
on process reliability and product quality.

Work Areas

Types of adhesives
Reactive adhesives
Dispersion adhesives
Vegetable adhesives, dextrin and starch
adhesives

Raw materials
Polymers

For applications in the field of
Wood/furniture industry
Construction industry, including floors, walls
and ceilings

ZHAW Laboratory of Adhesives

Laboratory of Adhesives and Polymer Materials
Zurich University of Applied Sciences (ZHAW)
Technikumstrasse 9, CH-8401 Winterthur
Phone +41 (0) 58 934 65 86
Email: christof.braendli@zhaw.ch
www.zhaw.ch/impe

Member of FKS

Company

Year of formation
School of Engineering: 1874, Institute: 2007

Size of workforce
School of Engineering: 580, Institute: 40

Managing partners
School of Engineering: Prof. Dr. Dirk Wilhelm,
Phone: +41 58 934 47 29,
Email: dirk.wilhelm@zhaw.ch
Institute of Materials and Process
Engineering: Prof. Dr. Andreas Amrein,
Phone: +41 58 934 73 51,
Email: andreas.amrein@zhaw.ch

Laboratory of Adhesives and Polymer
Materials. Prof. Dr. Christof Brändli,
Phone +41 58 934 65 86,
Email: christof.braendli@zhaw.ch

Ownership structure
Part of the Zurich University of Applied
Sciences (ZHAW)

Contact partners
Laboratory of Adhesives and Polymer
Materials: Prof. Dr. Christof Brändli,
Phone +41 58 934 65 86,
Email: christof.braendli@zhaw.ch

Further information
Applied R&D in the fields of materials,
adhesive bonding, and surfaces. Adhesive
development and testing. Broad range of
adhesive competencies including adhesive
chemistry and analysis.

Work Areas

Types of adhesives
Hot melt adhesives
Reactive adhesives
Solvent-based adhesives
Dispersion adhesives
Pressure-sensitive adhesives

Synthesis and Formulation
Adhesive formulation and synthesis
Polymer compounding, reactive extrusion,
and synthesis
Online reaction control with IR spectroscopy
Functionalization of nano particles

Characterization
Adhesive performance tests
Curing behavior studies
Thermal and mechanical analysis
Flow properties determination with
rheological methods
Morphological and surface analysis

Applications
Adhesive development
(structure-property relationship)
Polymer development
(incl. tapes, film, grafting, reactive, ...)
Nanomodifications

INSTITUTES AND RESEARCH FACILITIES

Research and Development

Adhesive bonding technology is a key contributor to the development of innovative products, creating the conditions for new, future-proof markets to open up across all industries. Small and medium-sized companies benefit from new joining methods by developing sophisticated products which give them competitive edge.

In order to profit from the advantages that adhesive bonding technology offers over other methods of joining, it is important to include the whole process from product planning through quality management to staff training.

This can only be achieved by close cooperation between research institutions and industry so that research results are applied quickly to the development of innovative products and production processes.

The following list includes all known research facilities and institutes which are committed to working with their partners in industry to resolve adhesive issues across a range of areas.

Deutsches Institut für Bautechnik
Abteilung II – Gesundheits- und Umweltschutz
(Federal Agency for the Approval of
Construction Products)
Kolonnenstraße 30 B
D-10829 Berlin

Contact:
Dipl.-Ing. Dirk Brandenburger MEM (UTS)
Phone: +49 (0) 30 78730 232
Fax: +49 (0) 30 78730 11232
Email: dbr@dibt.de
www.dibt.de

FH Aachen - University of Applied Sciences
Klebtechnisches Labor
Goethestraße 1
D-52064 Aachen

Contact:
Prof. Dr.-Ing. Markus Schleser
Phone: +49 (0) 241 6009 52385
Fax: +49 (0) 241 6009 52368
Email: schleser@fh-aachen.de
www.fh-aachen.de

Fogra Forschungsinstitut für
Medientechnologien e.V.
(Graphic Technology Research Association)
Einsteinring 1a
D-85609 Aschheim by München

Contact:
Dr. Eduard Neufeld
Phone: +49 (0) 89 43182 112
Fax: +49 (0) 89 43182 100
Email: info@fogra.org
www.fogra.org

FOSTA Forschungsvereinigung
Stahlanwendung e.V.
(Research Association for Steel Application)
Stahl-Zentrum
Sohnstraße 65
D-40237 Düsseldorf

Contact:
Dipl.-Ing. Rainer Salomon
Phone: +49 (0) 211 6707 853
Fax: +49 (0) 211 6707 840
Email: rainer.salomon@stahlforschung.de
www.stahlforschung.de

Fraunhofer-Institut für Fertigungstechnik und
Angewandte Materialforschung – IFAM
(Fraunhofer Institute for Manufacturing
Technology and Advanced Materials – IFAM)
Wiener Straße 12
D-28359 Bremen

Contact:
Prof. Dr. Bernd Mayer
Phone: +49 (0) 421 2246 419
Fax: +49 (0) 421 2246 774401
Email: bernd.mayer@ifam.fraunhofer.de
Prof. Dr. Andreas Groß
Phone: +49 (0) 421 2246 437
Fax: +49 (0) 421 2246 605
Email: andreas.gross@ifam.fraunhofer.de
www.ifam.fraunhofer.de

Fraunhofer-Institut für Holzforschung –
Wilhelm-Klauditz-Institut – WKI
(Fraunhofer Institute for Wood Research –
Wilhelm-Klauditz-Institute (WKI))
Bienroder Weg 54 E
D-38108 Braunschweig

Contact:
Dr. Heike Pecher
Phone: +49 (0)531 2155 206
Email: heike.pecher@wki.fraunhofer.de
www.wki.fraunhofer.de

Fraunhofer-Institut für Werkstoff- und
Strahltechnik – IWS
(Klebtechnikum an der TU Dresden, Institut
für Fertigungstechnik,Professur für Laser-
und Oberflächentechnik)
Winterbergstraße 28
D-01277 Dresden

Contact:
Annett Klotzbach
Phone: +49 (0) 351 83391 3235
Fax: +49 (0) 351 83391 3300
Email: annett.klotzbach@iws.fraunhofer.de
www.iws.fraunhofer.de

Fraunhofer-Institut für Zerstörungsfreie
Prüfverfahren – IZFP
(Fraunhofer Institute for Nondestructive
Testing – IZFP)
Campus E3.1
D-66123 Saarbrücken

Contact:
Prof. Dr.-Ing. Bernd Valeske
Phone: +49 (0) 681 9302 3610
Fax: +49 (0) 681 9302 11 3610
Email: bernd.valeske@izfp.fraunhofer.de
www.izfp.fraunhofer.de

Johann Heinrich von Thünen Institut (vTI)
Bundesforschungsistitut für Ländliche Räume,
Wald und Fischerei
Institut für Holzforschung
(Johann Heinrich von Thünen Institute (vTI)
Institute of Wood Research)
Leuschnerstraße 91
D-21031 Hamburg

Contact:
Dr. Johannes Welling
Phone: +49 (0) 40 73962 601
Fax: +49 (0) 40 73962 699
Email: hf@thuenen.de
www.thuenen.de/de/hf/

Hochschule München
Institut für Verfahrenstechnik Papier e.V. (IVP)
(Institute of Munich University of
Applied Science)
Schlederloh 15
D-82057 Icking

Contact:
Prof. Dr. Stephan Kleemann
Phone: +49 (0) 89 1265 1668
Fax: +49 (0) 89 1265 1560
Email: kleemann@hm.edu
www.hm.edu

Hochschule für nachhaltige Entwicklung
Eberswalde (FH)
Fachbereich Holzingenieurwesen
(Eberswalde University for Sustainable
Development
Department of Wood Engineering)
Schlickerstraße 5
D-16225 Eberswalde

Contact:
Prof. Dr.-Ing. Ulrich Schwarz
Phone: +49 (0) 3334 657 371
Fax: +49 (0) 3334 657 372
Email: ulrich.schwarz@hnee.de
www.hnee.de/holzingenieurwesen

IFF GmbH
Induktion, Fügetechnik, Fertigungstechnik
(Induction and Joining Technology Engineering)
Gutenbergstraße 6
D-85737 Ismaning

Contact:
Prof. Dr.-Ing. Christian Lammel
Phone: +49 (0) 89 9699 890
Fax: +49 (0) 89 9699 8929
Email: christian.lammel@iff-gmbh.de
www.iff-gmbh.de

ift Rosenheim GmbH
Institut für Fenstertechnik e.V.
(Institute for Window Technology)
Theodor-Gietl-Straße 7-9
D-83026 Rosenheim

Contact:
Prof. Ulrich Sieberath
Phone: +49 (0) 80 31261 0
Fax: +49 (0) 80 31261 290
Email: info@ift-rosenheim.de
www.ift-rosenheim.de

ihd – Institut für Holztechnologie Dresden GmbH
(Institute of Wood Technology)
Zellescher Weg 24
D-01217 Dresden

Contact:
Dr. rer. nat. Steffen Tobisch
Phone: +49 (0) 351 4662 257
Fax: +49 (0) 351 4662 211
Mobil: +49 (0) 1622 696330
Email: steffen.tobisch@ihd-dresden.de
www.ihd-dresden.de

Georg-August-Universität Göttingen
Institut für Holzbiologie und Holzprodukte
(Institute for Wood Biology and Wood
Technology)
Büsgenweg 4
D-37077 Göttingen

Contact:
Prof. Dr. Holger Militz
Phone: +49 (0) 551 393541
Fax: +49 (0) 551 399646
Email: hmilitz@gwdg.de
www.uni-goettingen.de

Institut für Fertigungstechnik
Professur für Fügetechnik und Montage
TU Dresden
(Institute of Manufacturing Technology, Joining
Technology and Assembly
Technical University of Dresden)
George-Bähr-Straße 3c
D-01069 Dresden

Contact:
Prof. Dr.-Ing. habil. Uwe Füssel
Phone: +49 (0) 351 46337 615
Fax: +49 (0) 351 46337 249
Email: uwe.fuessel@tu-dresden.de
https://tu-dresden.de/ing/
maschinenwesen/if/fue

IVLV - Industrievereinigung für Lebens-
mitteltechnologie und Verpackung e.V.
(Industry Association for Food Technology and
Packaging)
Giggenhauser Straße 35
D-85354 Freising

Contact:
Dr.-Ing. Tobias Voigt
Phone: +49 (0) 8161 491140
Fax: +49 (0) 8161 491142
Email: tobias.voigt@ivlv.org
www.ivlv.org

iwb – Anwenderzentrum Augsburg
Technische Universität München
(iwb – Augsburg Application Centre
Technical University of Munich)
Beim Glaspalast 5
D-86153 Augsburg

Contact:
Dipl.-Ing. Johannes Glasschröder
Phone: +49 (0) 821 56883 53
Fax: +49 (0) 821 56883 50
Email: johannes.glasschroeder@
iwb.mw.tum.de
www.iwb.mw.tum.de

Kompetenzzentrum Werkstoffe der Mikrotechnik
Universität Ulm
(WMtech University of Ulm)
Albert-Einstein-Allee 47
D-89081 Ulm

Contact:
Prof. Dr. Hans-Jörg Fecht
Phone: +49 (0) 731 50254 91
Fax: +49 (0) 731 50254 88
Email: info@wmtech.de
www.wmtech.de

Leibniz-Institut für Polymerforschung
Dresden e.V.
(Leibniz Institute of Polymer Research Dresden)
Hohe Straße 6
D-01069 Dresden

Contact:
Prof. Dr.-Ing. Udo Wagenknecht
Phone: +49 (0) 351 46583 61
Fax: +49 (0) 351 46583 62
Email: wagenknt@ipfdd.de
www.ipfdd.de

Naturwissenschaftliches und Medizinisches
Institut an der Universität Tübingen
(The Natural and Medical Sciences
Institute at the University of Tübingen)
Markwiesenstraße 55
D-72770 Reutlingen

Contact:
Dr. Hanna Hartmann
Phone: +49 (0)7121 51530 872
Fax: +49 (0)7121 51530 62
Email: hanna.hartmann@nmi.de
www.nmi.de

ofi Österreichisches Forschungsinstitut
für Chemie und Technik
Institut für Klebetechnik
(ofi Austrian Research Institute for Chemistry
and Technology
Institute for Adhesive Technology)
Franz-Grill-Straße 1, Objekt 207
A-1030 Wien

Contact:
Ing. Martin Tonnhofer
Phone: +43 (0)1798 16 01 201
Email: martin.tonnhofer@ofi.at
www.ofi.at

Papiertechnische Stiftung PTS
(Paper Technology Foundation)
Pirnaer Straße 37
D-01809 Heidenau

Contact:
Clemens Zotlöterer
Phone: +49 (0) 3529 551 60
Fax: +49 (0) 3529 551 899
Email: info@ptspaper.de
www.ptspaper.de

Prüf- und Forschungsinstitut Pirmasens e.V.
(Test and Research Institute for Footwear
Production)
Marie-Curie-Straße 19
D-66953 Pirmasens

Contact:
Dr. Kerstin Schulte
Phone: +49 (0)6331 2490 0
Fax: +49 (0)6331 2490 60
Email: info@pfi-germany.de
www.pfi-pirmasens.de

RWTH Aachen
ISF - Institut für Schweißtechnik und
Fügetechnik
(RWTH Aachen
ISF – Welding and Joining Institute)
Pontstraße 49
D-52062 Aachen

Contact:
Prof. Dr. -Ing. Uwe Reisgen
Phone: +49 (0) 241 80 93870
Fax: +49 (0) 241 80 92170
Email: office@isf.rwth-aachen.de
www.isf.rwth-aachen.de

Technische Universität Berlin
Fügetechnik und Beschichtungstechnik im
Institut für Werkzeugmaschinen und
Fabrikbetrieb
(Technical University Berlin
Joining and Coating Technology Departments
of Machine Tools and Factory Management)
Pascalstraße 8 – 9
D-10587 Berlin

Contact:
Prof. Dr.-Ing. habil. Christian Rupprecht
Phone: +49 (0) 30 314 25176
Email: info@fbt.tu-berlin.de
www.fbt.tu-berlin.de

Technische Universität Braunschweig
Institut für Füge- und Schweißtechnik
(Technical University Braunschweig
Institute of Joining and Welding Technology)
Langer Kamp 8
D-38106 Braunschweig

Contact:
Univ.-Prof. Dr.-Ing. Prof. h.c. Klaus Dilger
Phone: +49 (0) 531 391 95500
Fax: +49 (0) 531 391 95599
Email: k.dilger@tu-braunschweig.de
www.ifs.tu-braunschweig.de

Technische Universität Kaiserslautern
Fachbereich Maschinenbau und
Verfahrenstechnik
Arbeitsgruppe Werkstoff- und Oberflächen-
technik Kaiserslautern (AWOK)
(Technical University Kaiserslautern
Work Group for Materials and Surface
Technologies)
Gebäude 58, Raum 462
Erwin-Schrödinger-Straße
D-67663 Kaiserslautern

Contact:
Univ.-Prof. Dr.-Ing. Paul Ludwig Geiß
Phone: +49 (0) 631 205 4117
Fax: +49 (0) 631 205 3908
Email: geiss@mv.uni-kl.de
www.mv.uni-kl.de/awok

TechnologieCentrum Kleben
TC-Kleben GmbH
(Center Adhesive Bonding Technology)
Carlstraße 50
D-52531 Übach-Palenberg

Contact:
Dipl.-Ing. Julian Brand
Phone: +49 (0) 2451 9712 00
Fax: +49 (0) 2451 9712 10
Email: post@tc-kleben.de
www.tc-kleben.de

Universität Kassel
Institut für Werkstofftechnik, Kunststofffüge-
techniken, Werkstoffverbunde
(University of Kassel
Institute für Materials Engineering)
Mönchebergerstraße 3
D-34125 Kassel

Contact:
Prof. Dr.-Ing. H.-P. Heim
Phone: +49 (0) 561 80436 70
Fax: +49 (0) 561 80436 72
Email: heim@uni-kassel.de
www.uni-kassel.de/maschinenbau

Universität Kassel
Fachgebiet Trennende und Fügende Fertigung-
verfahren
(University of Kassel
Department of Separative and Joining
Production Processes)
Kurt-Wolters-Straße 3
D-34125 Kassel

Contact:
Prof. Dr.-Ing. Prof. h.c. Stefan Böhm
Phone: +49 (0) 561804 3236
Fax: +49 (0) 561804 2045
Email: s.boehm@uni-kassel.de
www.tff-kassel.de

Universität des Saarlandes
Adhäsion und Interphasen in Polymeren
(University of Saarland
Adhesion and Interphases in Polymers)
Campus, Geb. C6.3
D-66123 Saarbrücken

Contact:
Prof. rer. nat. Wulff Possart
Phone: +49 (0) 681 302 3761
Fax: +49 (0) 681 302 4960
Email: w.possart@mx.uni-saarland.de
www.uni-saarland.de/nc/startseite

Universität Paderborn
Laboratorium für Werkstoff- und Fügetechnik
(University of Paderborn
Laboratory for Materials and Joining Technology)
Pohlweg 47-49
D-33098 Paderborn

Contact:
Prof. Dr.-Ing. Gerson Meschut
Phone: +49 (0) 5251 603031
Fax: +49 (0) 5251 603239
Email: meschut@lwf.upb.de
www.lwf-paderborn.de

Wehrwissenschaftliches Institut für Werk- und
Betriebsstoffe - WIWeB
(Bundeswehr Research Institute for Materials,
Explosives, Fuels and Lubricants – WIWeB)
Institutsweg 1
D-85435 Erding

Contact:
Phone: +49 (0) 81 229590 0
Fax: +49 (0) 81 229590 3902
Email: wiweb@bundeswehr.org
www.baainbw.de/portal/a/baain/start/
diensts/wiweb

Westfälische Hochschule Abteilung
Recklinghausen
Fachbereich Ingenieur- und
Naturwissenschaften
Organische Chemie und Polymere
(Westfälische Hochschule
Department of Industrial Engineering,
Organic Chemistry and Polymers)
August-Schmidt-Ring 10
D-45665 Recklinghausen

Contact:
Prof. Dr. Klaus-Uwe Koch
Phone: +49 (0) 2361 915 456
Fax: +49 (0) 2361 915 751
Email: klaus-uwe.koch@w-hs.de
www.w-hs.de/erkunden/fachbereiche/
ingenieur-und-naturwissenschaften/
portrait-des-fachbereichs/

German
Adhesives
Association
Industrieverband Klebstoffe e.V.

ANNUAL REPORT 2018

Economic Report

The German adhesives industry continued to grow in 2017 and 2018. The industry's domestic revenues in 2017 from all types of adhesive systems, including adhesives, sealants, cement-based systems and adhesive tapes, amounted to € 3.85 billion. This equates to a growth rate of 2.7 percent. The positive economic situation of the construction industry was a significant driver of growth in this area.

Following an increase in sales and stable economic conditions during the first half of 2018, in the third quarter the market began to contract and in the fourth quarter the industry saw an economic downturn. The reason for the decline in sales in the second half of 2018 was a drop in the demand for adhesives in particular from the automotive and electronics industries.

Against this background and taking into consideration the various macroeconomic market indicators and the production statistics supplied by the Federal Statistical Office of Germany, the German adhesives industry saw overall average growth of 2% in nominal terms in 2018 and, in addition, positive developments on the export side. The total market in 2018 amounted to around € 3.95 billion.

In 2018 the German adhesives industry continued to face challenges, which included the availability of key raw materials, exchange rate effects, a shortage of skilled employees and transport capacity.

The German adhesives industry is in a very strong position on European and other international markets. With a global market share of more than 19 percent, it is the world market leader. It also holds the top positions on the European market, with an adhesives consumption of 27 percent and a production share of more than 34 percent.

Worldwide, sales revenues of approximately € 61 billion per year (not adjusted for exchange rate effects) were generated with adhesives, sealants, adhesive tapes and system products. The German adhesives industry, which consists mainly of medium-sized companies, plays a prominent role on the international stage. Most of the companies manufacture their products in Germany and export them worldwide. In addition, some German adhesives firms supply the world markets from more than 200 local manufacturing plants outside Germany.

With both business models, the German adhesives industry generates sales revenues of € 11.9 billion worldwide. Exports from Germany amount to more than € 1.7 billion, while German adhesives manufacturers generate a further € 8.1 billion of sales revenues locally from their production facilities outside Germany. The German market has a sales volume of almost € 4 billion per year. As a result of the use of "adhesive systems developed in Germany" in almost all manufacturing industries and in the construction sector, the adhesives industry was responsible for indirect value creation of significantly more than € 400 billion in Germany. Worldwide, value creation amounted to more than € 1 trillion.

This strong position is the direct result of innovative technological developments for the paper/packaging, automotive, wood, electronics and construction industries, where the adhesives industry provides practical, value-added solutions.

The link between the price of crude oil and of basic raw materials for adhesives weakens

The price of crude oil is no longer directly reflected in the price of end products. This connection has almost been broken. Prices are increasingly being determined by the many refining processes involved in the value chain (see the graphic "Raw material flows") and the availability of a raw material on the market. The link between the price of a range of basic raw materials and that of crude oil is likely to weaken even further. This is made very clear by a comparison between the fluctuations in the price of crude oil and of acetic acid, VAM, ethylene and methanol (see the graphic "Price fluctuations in crude oil and basic raw materials"). While the price of crude oil fell throughout 2015 and 2016 for example, the prices of the raw materials referred to above remained largely stable or were heavily influenced by regional supply and demand. In particular in the case of high-quality adhesives, the fluctuations in prices are no longer linked because there can be up to 10 processing stages between crude oil and the raw materials used to manufacture PUR, epoxy resin and acrylate adhesives.

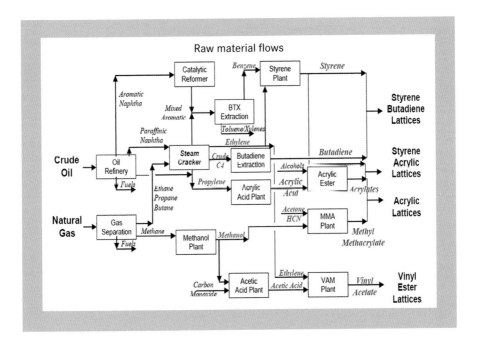

Raw material flows

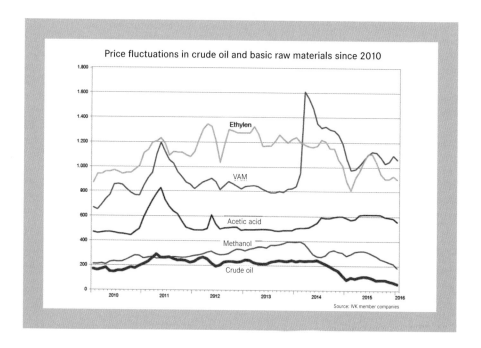

Price fluctuations in crude oil and basic raw materials since 2010

Source: IVK member companies

In the summer of 2018 significant increases in the prices of basic chemicals and intermediate products combined with limited availability of these materials and a strong global demand led to considerable rises in the costs of adhesive raw materials. In addition, the constantly rising costs imposed by regulations and a shortage of freight capacity in Europe resulted in a noticeable increase in pressure on the cost structure of the German adhesives industry.

About our Committee Work

General Assembly

The general assembly meetings in 2018 in Stuttgart and in 2019 in Cologne took place in the context of a deterioration in the overall economic situation.

Following an increase in sales and stable economic conditions during the first half of 2018, in the third quarter the market began to contract and in the fourth quarter the adhesives industry saw an economic downturn. The reason for the decline in sales in the second half of 2018 was a drop in the demand for adhesives, in particular from the automotive and electronics industries.

Against this background and taking into consideration the various macroeconomic market indicators (forecasts for customer industries, GDP and IPX) and the production statistics supplied by the Federal Statistical Office of Germany, the German adhesives industry saw overall average growth of 3% in nominal terms in 2018 and, in addition, positive developments on the export side. The total German market in 2018 for all adhesives, sealants, construction adhesives and adhesive tapes amounted to around € 4 billion.

In 2018 the German adhesives industry continued to face challenges, which included the availability of key raw materials, exchange rate effects, a shortage of skilled employees and transport capacity.

As a result of ongoing geopolitical risks (such as the escalation of trade disputes, further increases in indebtedness, the Brexit chaos etc.), the first 5 months of 2019 have seen an ongoing deterioration in the economic growth of some customer sectors of the adhesives industry. Only the construction sector is still in a strong position.

Although more stable exchange rates and good availability of raw materials are currently having a stabilising effect on the economic situation of the German adhesives industry, the mood in the industry for the financial year 2019 is generally less optimistic than it was a year ago.

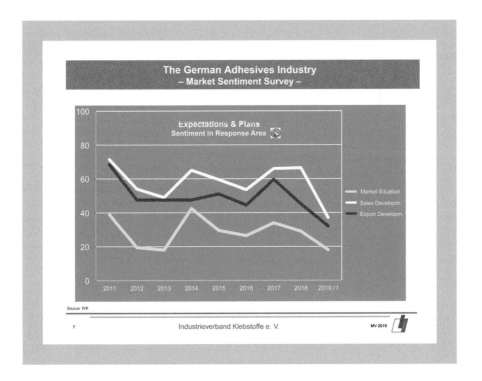

The services that the German Adhesives Association (IVK) provides for its members cover three areas:
- Technology
- Communication
- Future initiatives

IVK continues to give the highest priority to **technical subjects** because working together to manage and implement the legislative requirements in particular that apply to adhesives companies has a significant impact on the success of the industry. This applies to REACH, labelling (GHS/CLP), permitted biocides, (mandated) standardisation and sustainability, as well as technical briefing notes, seminars and conferences and contacts with customer organisations.

The subject of the **circular economy** is increasingly becoming a major socio-political trend. The EU Ecodesign Directive, which originally focused on energy efficiency, has now been extended to cover the material efficiency of energy-related products. In order to clarify the ecodesign requirements, a number of implementing regulations have been introduced. These included "no gluing" requirements which caused some alarm within IVK.

Together with FEICA, the German Adhesives Association has had meetings on a European and a national level and has successfully pressed for a wording concerning the requirements for adhesives that does not specify a particular technology.

The EU Ecodesign Directive and the implementing regulations are just one part of a larger-scale EU circular economy strategy, which is currently the focus of discussions within the EU. As part of the development of the EU framework legislation, a list has recently been published of the high-priority areas that the future regulatory activities will concentrate on. These areas are as follows:
- Plastics
- Food/food waste
- Critical raw materials
- Construction and demolition waste
- Biomass
- Bio-based products

Adhesives play a role in many of these fields, so there is a risk that a highly critical approach will be taken to bonding in this context and that it will be governed by strict regulations. The discussions with the authorities have shown that the industry would be very well advised to significantly expand its strategy for communicating with the technical departments of the ministries and with the associated institutions. More and more (scientific) studies are being initiated by the authorities which take a highly critical approach to adhesives and bonding in relation to the circular economy.

IVK has set up an advisory board that will look in detail at the subject of the circular economy and ecodesign from the perspective of the adhesives industry. In addition, the association has commissioned a scientific study in order to demonstrate that bonding is not an obstacle to the circular economy, but instead a part of the solution. Experience in this area shows that the results of scientific studies are more readily accepted and used for decision-making purposes by ministries and other public bodies than simple industry information.

Sample EPDs (sample environmental product declarations for construction adhesives) and guidelines are available on the IVK website in German and English at http://epd.klebstoffe.com/. These have now been certified by IBU (the German institute for construction and the environment) with the help of the consulting company Thinkstep on the basis of European data and published by FEICA (the Association of the European Adhesive and Sealant Industry) at http://www.feica.eu/our-priorities/key-projects/epds.aspx. They are widely recognised in Europe, but in some cases additional national data is required. During 2018 and 2019, the European sample EPDs will replace the German EPDs, in order to keep the work involved in updating them to a minimum. The EU Commission is developing EN 15804, the basic standard for EPDs, for the purposes of harmonisation. It plans to add toxic criteria to the list of impact indicators, such as the greenhouse effect and the potential for ozone depletion, acidification and eutrophication, which are used for the creation of life cycle assessments.

In addition, IVK and its committees are actively working on a number of different issues and projects as part of the association's portfolio of services, which is based on the needs of its members. These include:
- Close involvement in the standards process on a national and a European level
- The introduction of the DIN 2304 standard "Quality requirements for adhesive bonding processes"
- Support by EMICODE for very low emission products as part of the IVK sustainability concept
- Publication of information on the approval of various in-can preservatives, the resulting changes in labelling requirements and the accompanying deadlines

A number of activities have been taking place in the field of communication.
- A new issue of the magazine **"Kleben fürs Leben"** was published to coincide with the general assembly meetings. Since mid March 2018, "Kleben fürs Leben" has also been issued as an **online magazine.** The magazine has been available in print form for more than 10 years and has been accompanied by an e-paper for the last five years. An independent online version has now been created. Every month it will contain interesting multimedia articles from the world of adhesives which will include pictures, videos and audio material alongside the text. The articles from previous issues of the print magazine are also being digitised and more than 100 are already available online.

- Regular press releases have been issued that have received a **response from a wide variety of media organizations.** Many articles on the subject of bonding have been published in the print and online media, on the IVK press portal www.klebstoff-presse.com and on social media networks.

- The process of **digitising the IVK's educational materials,** linking them to other information provided by the association (job profiles for the adhesives industry, the promotional video, the adhesives guidelines) and adding information about the chemistry of adhesives has now been completed. This gives education specialists a tool that meets the new requirements for digital education. Alongside the digitisation process, the educational materials have been revised and expanded by the University of Frankfurt specifically for the purposes of teacher training.
- The Technical Committees have produced a total of **60 publications** (technical briefing notes, reports, information sheets). These are an important and much valued source of information for the customers of the adhesives industry. Almost all of them are also available in English.
- Conferences and seminars have been held for members and their customers on a range of different technical and regulatory subjects. These events are also an ideal opportunity for networking, meeting existing contacts and making new ones.

As the official production statistics no longer provide reliable information, the executive board has decided to develop internal production statistics for the association which have no implications with regard to competition law and involve very little work for the member companies. Data such as net external sales and quantities and areas in tonnes and m^2 will be generated for adhesives, sealants, cement-based products and adhesive tapes. In addition, the companies will report on domestic sales, exports, sales by foreign subsidiaries of German firms and global sales, together with the number of employees and production facilities. An agency will be responsible for generating the data.

During the 2018 and 2019 general assemblies, the following companies were accepted as members of the German Adhesives Association:

- BCD Chemie GmbH, Hamburg
- Bilgram Chemie GmbH, Ostrach
- Fenos AG, Freiberg a.N.
- Grünig KG Professional Adhesives, Bad Kissingen
- L&L Products Europe GmbH, Kehl
- Möller Chemie GmbH & Co. KG, Steinfurt
- PCC Specialties GmbH, Kamp-Lintfort
- Perstorp Service GmbH, Arnsberg
- Plasmatreat GmbH, Steinhagen
- SCIGRIP Europe, Washington, UK
- Süddeutsches Kunststoffzentrum, Würzburg
- Wöllner GmbH, Ludwigshafen

Sustainability

Sustainability means preserving a system for the good of future generations. Doing business sustainably does not mean focusing on the short-term success of a product or service, but instead concentrating on making a positive contribution in the long term. In addition, environmental, social and economic factors need to be taken into consideration.

Many lengthy discussions have been held by IVK as to whether and how sustainability factors can create added value for society and the economy without taking action just to be seen to be doing so and without incurring unnecessary costs. However, now that politicians and standardisation committees have taken up the topic, the future direction has been set. Even in some individual sectors of the economy, for example large-scale building projects, proof of compliance with "sustainable" building practices is occasionally required. The objective of resource-efficient building is specified in the Construction Products Regulation EU/305/2011 (CPR-BRCW7).

The objective of sustainable building involves the comprehensive analysis and optimisation of buildings to improve the quality of the environment, society and the economy throughout the entire life cycle of the structures. Currently, the focus is on assessing and evaluating environmental aspects of energy and resource consumption. A wide variety of organisations have already taken measures to determine how sustainability can be verified and certified. In Germany, DGNB is responsible for the certification of buildings, while BNB assumes responsibility for government buildings. In other countries, HQE, BREEAM, LEED and other systems have been established. Since the sustainability impact assessment of buildings is primarily based on the products used and their upstream products and production processes, these also need to be evaluated. In the case of construction products, the consultancy company Thinkstep has the information needed (for example, data on primary energy consumption and impact indicators, such as the greenhouse effect, ozone depletion, potential acidification and eutrophication) to draw up life cycle assessments and carry out life cycle analyses, which are an integral part of the **EPDs (Environmental Product Declarations)** in accordance with EN 15804. In Germany and in other countries, the Institut Bauen und Umwelt e. V. (Institute for Construction and Environment (IBU)) plays a major role in developing and issuing EPDs. IBU has an EPD programme that includes preparing EPDs, reviewing them for completeness and plausibility and having them verified by an independent third party. IVK is a member of IBU and has been working closely with its sister associations Deutsche Bauchemie (German Industry Association for Construction Chemicals (DBC)) and Verband der deutschen Lackindustrie (Association of the German Paint Industry (VdL)) since 2010 to develop an industry solution for the creation of sample EPDs, initially for the German market, which are intended to help manufacturers to provide the necessary information for building certification. Sample EPDs have been drawn up for a wide variety of building products, including construction adhesives, which cover almost the entire industry. In order to meet market requirements, this commitment has been extended to cover the whole of Europe and, as a result, European sample EPDs have now been made available with the support of FEICA and the assistance of EFCC and can be downloaded by the members of the relevant associations from http://www.feica.eu/our-priorities/key-projects/epds.aspx.

Prominent certification bodies are supporting the project, which is also being promoted by ECO Platform (http://www.eco-platform.org/) using its logo. The members of ECO Platform are the leading European program operators. The aim is to meet the key requirement of the Construction Products Regulation concerning sustainable building (BRCW7).

The **"(Product) Carbon Footprint" (PCF)** is more suitable for industrial use. This is an attempt to reduce the sustainability assessment to a single value – the CO_2 emissions of a product. This approach, which was in part initiated by politicians, has since been rejected by many experts. Preparing a PCF for adhesives, which are semi-finished products, makes little sense, because they are never used on their own but always as a component of a system with at most a marginal impact on the PCF of the final product. A project that was organised jointly with the Bundesverband Druck & Medien (German Print and Media Association) to assess the impact of adhesives on the carbon footprint of the production of books clearly shows that the percentage of the CO_2 emissions that can be attributed to the adhesive is between 0.06 and 0.5%, depending on the type of adhesive used (dispersion, hot melt), and is therefore negligible when compared to the use of paper. It can therefore be assumed that the carbon footprint of adhesives is very small compared to the products that are manufactured using them. Although the quantity of adhesive used in the manufacturing of products is usually very small, adhesive manufacturers are also asked about the PCF values of the adhesives that they supply. Many studies in recent years have shown that the PCF value of adhesives in their delivered state (cradle-to-gate) can be summarised within certain product groups, taking into consideration the error thresholds of PCF calculations. On the initiative of the Technical Board, PCF values (cradle-to-gate) have been calculated for the following product groups:

- Adhesives based on aqueous plastic dispersions
- Thermoplastic hot melt adhesives
- Solvent-based adhesives
- Reactive adhesives

The corresponding PCF documents and more detailed information on the individual product groups and the database can be found in the IVK information sheet "Typical Product Carbon Footprint (PCF) values for industrial adhesives". (http://www.klebstoffe.com/fileadmin/redaktion/ivk/M-RS_2014-19_Anl_Product_Carbon_Footprint___www__.pdf).

Demonstrating that adhesive bonding produces more sustainable product solutions than alternative joining methods is much more important than providing the PCF for an adhesive bonding system and confirming its marginal impact on the carbon footprint of the end product.

Resource-efficient lightweight design in the automotive industry is an outstanding example of the positive influence of adhesives. Using epoxy resins to produce rotor blades for wind turbines is another example and clearly demonstrates how the carbon footprint of a product can be improved by using modern materials technology. This opens up new market opportunities because it allows sustainable solutions for selected applications to be highlighted and eliminates the need for individual analyses. You can find other examples of this at http://www.feica.eu/our-priorities/sustainable-development.aspx.

The subject of the circular economy, which is being promoted by the EU, and the related action plan entitled "Closing the loop – An EU action plan for the Circular Economy" are growing in importance throughout Europe. The EU's activities in this area include its plastics strategy, bioeconomy strategy, waste package and Construction Products Regulation, all of which could have an effect on the European adhesives industry.

The first direct impact on the adhesives industry was the result of the requirements specified in the Ecodesign Directive 2009/125/EC concerning the environmentally friendly design of energy-related products. Alongside energy efficiency, the European Commission is also focusing on factors such as resource and material efficiency, durability, repairability and reusability. Washing machines, computers/servers and displays are the product groups currently affected and others will soon be included.

The adoption of implementing regulations allows the EU Commission to introduce product regulations without the involvement of the European Parliament. To provide support for this process, the EU Commission issued the standardisation mandate M/543 "Ecodesign requirements on material efficiency aspects for energy-related products" to CEN, CENELEC and ETSI. Horizontal standards will be drawn up covering issues such as the recyclability, recoverability, reusability, disassembly and service life of energy-related products. These horizontal standards will form the basis for product-specific standards.

In discussions with public bodies, in a variety of public consultations and in a position paper, IVK has made it clear that:

- Bonding does not prevent a product from being repaired or recycled. (However, debonding must be one of the requirements placed on product manufacturers and must be taken into consideration during the product design process, in collaboration with the adhesives supplier.)
- Only a formulation which is technology-neutral (the specification of objectives instead of joining methods) will allow innovations to be developed in future.

A proposal for restrictions on the use of microplastics as part of REACH was published in January 2019 and revised in March 2019. It has yet to be finalised, but it includes notification and reporting requirements that could affect some types of adhesives. Working closely with VCI and other industry associations, IVK has used the public consultation to highlight the disproportionate impact of the reporting requirements, in particular on small and medium-sized businesses.

The initiatives launched by a range of companies, industries and organisations that use adhesives also need to be monitored.

In order to provide expert technical support in all these areas in future, IVK established an advisory board for the circular economy in 2018. At the suggestion of the Executive Board and with the approval of the general assembly, IVK commissioned the Fraunhofer IFAM to carry out a scientific study entitled "The circular economy and adhesive technology" and provided a budget of €100,000 for the project. The study is being supported by the Technical Board and the advisory board for the circular economy.

The IVK management team believes that the circular economy and ecodesign represent a new regulatory area that could have a major influence on the requirements placed on adhesive bonding in the decades to come.

Guidelines "Adhesive Bonding – the Right Way"

Creating a strong bonded joint involves more than just choosing the right adhesive. Other important factors that need to be taken into consideration include the properties of the materials, the surface treatment, a design suitable for bonding and proof of the reliability of the joint in use. The German Adhesives Association has collaborated with the Fraunhofer Institute IFAM to develop the interactive guidelines "Adhesive bonding – the right way". They are intended for tradespeople and industrial businesses that need additional information about adhesive systems.

Adhesives now have such a wide variety of uses that it is not possible for adhesives manufacturers to cover all the possible applications and, in particular, highly specialised ones in their data sheets. The guidelines "Adhesive bonding – the right way", which have been created by the German Adhesives Association and the Fraunhofer Institute IFAM, are a practical guide for tradespeople and industrial companies that need basic or more advanced information. The process of designing, developing and manufacturing an imaginary product is explained step by step and all the stages in the planning and manufacturing process are covered systematically. Carrying out all the necessary steps in the correct order makes a significant contribution to the quality of the end product. The interactive guidelines are the ideal tool for achieving this and they also include a glossary and search function. They cover the most important areas in the practical use of adhesives technology.

The English version of the guidelines is available free of charge in interactive form on the web portal of the German Adhesives Association: http://onlineguide.klebstoffe.com

Executive Board

The membership of the Executive Board of the German Adhesives Association reflects the existing corporate structure of the German adhesives industry, which consists of small and medium-sized businesses as well as companies operating at a multinational level. Furthermore, the balance in the board's members guarantees access to the highest level of core competence and expertise relating to the key market segments that are important to the adhesives industry.

A top priority for the Executive Board of the German Adhesives Association is to adapt the association's structure and its committees to new economic and technological constraints and conditions quickly and on an ongoing basis, in order to ensure that the organisation always operates efficiently and provides the maximum benefit for the German adhesives industry.

Holding technical discussions, assessing the economic, political and technological trends in the various key market segments of the adhesives industry and monitoring and analysing the activities of the association's numerous commercial and technical committees form an integral part of the responsibilities of the association's Executive Board.

The German Adhesives Association is considered to be the foremost competence centre in the field of adhesive bonding and sealing. It is the world's largest national association and the leading body in terms of its comprehensive service portfolio for adhesive bonding technology. IVK's links with the German Chemical Industry Association (VCI) and its various technical sections form the basis for its successful position. In addition to its connections with the chemicals industry, the association has a strategic and highly-efficient 360-degree network of expertise consisting of all the relevant system partners, scientific institutions, leading trade and industry associations, employers' liability insurance associations, consumer organisations and the organisers of exhibitions, training courses and conventions. As a result, it covers every single element in the adhesives value chain.

Within the framework of the strategy developed by the association's Executive Board for a "qualified market expansion", IVK actively supports a wide variety of scientific research projects relating to the field of adhesive bonding technology. This systematic approach to research is primarily interdisciplinary. In practical terms, this means that scientific knowledge is combined with engineering expertise in order to produce practical research results. Consequently, this approach has contributed to the fact that adhesive bonding has now become predictable and is firmly established as a reliable bonding and sealing technology in the field of engineering. It is now rightly and undisputedly considered to be the key technology of the 21st century.

As a founding member of the ProcessNet adhesives group under the umbrella of DECHEMA and the joint committee on adhesive bonding (GAK), IVK maintains regular contact with all the relevant research institutions involved in steel, timber and automotive research. Together, they jointly assess and support publicly funded scientific research projects in the field of adhesive bonding technology. This cooperation has enabled the German Adhesives Association to successfully establish important research projects with results that promise significant benefits in particular for the adhesives companies which are members of the association. The approved project concepts and the associated results are regularly published on a joint website and presented during the annual DECHEMA colloquium "Gemeinsame Forschung in der Klebtechnik" (Joint research in adhesives and bonding technology).

The employee training programme of the Fraunhofer IFAM in Bremen, which has received strong support from IVK in terms of finance and content, has evolved into a well-established, recognised educational programme. More and more companies that use adhesives have realised the measurable advantages of providing their employees with training in the appropriate use of technically demanding adhesive systems. For example, bonded joints used in the manufacture of rail vehicles are now produced only by trained staff. The adhesives industry itself benefits in every respect from expert system partners.

Alongside the employee training programme, a DIN standard to ensure the quality of load-bearing (structural) bonded joints used in specific safety areas has been developed and published with the support of the association's technical committees. This standard is currently being transferred to the international arena as part of an ISO standards project.

The Executive Board's strategy to extend its employee training programme to the European and international markets has been successful. The course contents, which were developed by the Fraunhofer IFAM for the various levels of qualifications (adhesive bonder, adhesive specialist, adhesive engineer) with the financial support of the association, have been translated into English, Chinese and other languages and adapted to the different applicable European and international standards. Training courses for adhesive bonders are now held on a regular basis in Poland, the Czech Republic, Turkey, the United States, China and South Africa. More than 10,000 adhesive bonders, specialists and engineers throughout the world have now been trained by the Fraunhofer IFAM team based in Bremen.

By developing this global training standard and putting in place a suitable worldwide training programme, the Executive Board of the German Adhesives Association has once again clearly demonstrated the key position of Germany's adhesives industry on international markets.

For IVK's Executive Board, providing sound training and qualifications in adhesive bonding technology is just as important as ensuring that young people have a good education in subjects such as chemistry, engineering and material sciences. Against this background, the contents and educational approach of the programme "Die Kunst des Klebens" (The Art of Bonding) have been comprehensively revised by experts from IVK, the German Chemical Industry Fund (FCI) and educational specialists. The course material is available to almost 20,000 technology teachers all over Germany.

Against the background of the initiative to digitise education in schools which has been adopted and launched by the German federal government, the Executive Board has decided to digitise its educational programme "Die Kunst des Klebens" (The Art of Bonding) and to link it to other information material provided by the association (the e-paper "Job Profiles", the promotional video, the adhesives guidelines, the adhesives magazine etc.). The association's partner for this project is the educational publisher Hagemann Bildungsmedienverlag, which used its technical expertise to digitise the educational material. As a result, teachers will have a tool that meets the new requirements for digital vocational education.

The adhesives industry is one of the first sectors of the chemical industry to offer teachers a digital package of this kind. This increases the chance of the subject of bonding and adhesives being included in the school curriculum. The digital teaching material was presented for the first time at the didacta 2018 fair in Hanover and received a positive response from the education profession. It was officially made available to schools at the start of the school year 2018/2019.

In addition to this teaching material, the German Adhesives Association, in cooperation with FWU, the Institute for Film and Picture in Science and Education, has designed and produced two educational DVDs. The DVD entitled "Grundlagen des Klebens" (Adhesive Fundamentals) was developed for classroom use in schools and vocational colleges, while "Kleben in Industrie und Handwerk" (Adhesives in Industry and Trades) describes specific practical applications of adhesives involving different combinations of materials and is ideal for use in engineering and materials science teaching at vocational colleges. Both DVDs consist of a variety of films, animated sequences, interactive assessment tests and extensive information for instructors and students. The DVDs are available online from the FWU media library at www.fwu.de.

The subject of adhesives and bonding in teaching is covered in greater depth during the regular seminars for teachers held by the German Chemical Industry Association (VCI).

The association is also supporting Germany's unique, nationwide development program for young people known as "Fraunhofer MINT-EC Talents", which aims to provide support for particularly gifted school students and to encourage them to study the STEM subjects (science, technology, engineering and mathematics). Scientists from the Fraunhofer IFAM run workshops on the subject of bonding for talented young people over a period of two years until they leave secondary school. The high levels of participation in the "Jugend forscht" research competition for young people and the excellent results it has produced clearly demonstrate the benefits of providing support at an early stage.

IVK is also increasingly becoming a leading player both in Europe and worldwide with regard to technical issues.

This applies in particular to the concept, jointly developed by the Executive Board and the Technical Board, of a standardisation competence platform in the German Adhesives Association. The objective of this project is to make use of the in-depth involvement of the German Adhesives Association in European standardisation (CEN) and to play an active role in international standardisation activities at the ISO level. The main factors driving this initiative were, on the one hand, the increasing number of ISO standards for adhesives which are gaining growing acceptance as part of the globalisation of markets. On the other hand, the Executive Board is pursuing the long-term goal of ensuring that German industrial standards which are important for the adhesives industry become established globally. With its standardisation competence platform, the association has succeeded in setting up and implementing important standards projects in the fields of adhesives for floor coverings and wood on an international level. In addition, it has provided key information to member companies about future developments in the electronics industry and the requirements for the relevant adhesives.

The German Adhesives Association has taken responsibility for managing the European standardisation project "Mandated flooring adhesive standard" and the European Standards Secretariat "Wood and Wood Materials", as a result of the lack of any obvious interest in these areas on the part of FEICA, the European umbrella organisation. These projects will ensure that the interests of Germany's manufacturers of construction adhesives have adequate representation with respect to European requirements on indoor air pollution, with the aim in the medium term of avoiding the need to comply with the emission testing regulations of the German Institute for Building Technology (DIBt), which involve a considerable amount of bureaucracy.

By leading the Standards Secretariat for Wood and Wood Materials, the association will be able to protect the specific interests of the German adhesives industry in the complex regulated market for glued laminated timber for load-bearing applications.

The subject of the circular economy, which is being promoted by the EU, and the related action plan entitled "Closing the loop – An EU action plan for the Circular Economy" are growing in importance throughout Europe. The EU's activities in this area include its plastics strategy, bioeconomy strategy, waste package and Construction Products Regulation. The first direct impact of this on the adhesives industry was the result of the requirements specified in the Ecodesign Directive concerning the environmentally friendly design of energy-related products. These were quickly followed by the proposal for restrictions on the use of microplastics as part of REACH. This proposal, which still has to be finalised, includes notification and reporting requirements that could affect some types of adhesives. The initiatives launched by a range of companies, industries and organisations that use adhesives also need to be monitored. In order to provide expert technical support in all these areas in future, the German Adhesives Association established an advisory board for the circular economy in 2018. At the suggestion of the Executive Board and with the approval of the general assembly, IVK commissioned the Fraunhofer IFAM to carry out a scientific study entitled "The circular economy and adhesive technology" and provided a budget of € 100,000 for the project.

In order to be able to take a practical approach to sustainability, a subject currently being given a high priority by politicians and the business world, the Executive Board has set up a new platform which will focus on two main areas:

- In cooperation with the German Industry Association for Construction Chemicals and the German Paint and Printing Ink Association, an industry solution will be drawn up for generic environmental product declarations (EPDs), which have been required since 2011 under the terms of the EU Construction Products Regulation as proof of the sustainability of construction products. These EPDs include detailed analyses and documentation (CO_2 emissions in production, energy consumption, environmental assessments, life cycle analyses etc.). They comply with the requirements of standards, create a uniform framework for the industry and can be used by all the relevant members of the association. Companies can upgrade individual EPDs at any time.
- The sample EPDs attracted a great deal of interest throughout Europe, which is an important market for German manufacturers of construction adhesives. For this reason, they have been taken over by FEICA with the technical and financial support of IVK and have been recognised across Europe. As a result, the sample EPDs relating only to the German industry have become less important and will not be continued after 2020 when they expire. However, the revision of EN 15804, the relevant standard in this respect, requires the EPDs, which have to be reviewed every five years, to provide information about end-of-life indicators. This has been put into effect as part of an industry-wide FEICA project and provides guidance for the future implementation of the basic requirement for sustainability laid down by the EU Construction Products Regulation.
- A parallel project that evolved from the EPD project has focused on calculating carbon footprints for individual families of industrial adhesives. Information about CO_2 emissions (carbon footprint) was originally required by the food industry, but is now increasingly being called for by other industries that use adhesives.

The Executive Board made available a budget of over € 250,000 for these projects (EPDs, carbon footprint, European and US adaptation).

A project that was organised jointly with the Bundesverband Druck & Medien (German Print and Media Association) to assess the impact of adhesives on CO_2 emissions during the production of books clearly shows that the percentage of CO_2 that can be attributed to the adhesive is between 0.06 and 0.5%, depending on the type of adhesive used, and is therefore negligible.

As a result, it can generally be assumed that the CO_2 footprint of adhesives has only a marginal impact on CO_2 emissions during the production of glued products. Additional practical examples will be collected on an ongoing basis to substantiate this assumption.

Regardless of the results of these additional investigations, it is clear that innovative, modern adhesive systems are essential for producing eco-efficient products. This applies both to light-weight construction in the automotive and furniture industries and to the manufacturing of products used to generate renewable energies, such as solar cells and wind turbines.

The subject of the circular economy is increasingly becoming a major new social trend that goes beyond consideration of the valuable contribution made by adhesives to the manufacture of sustainable and environmentally efficient products. In the first EU Ecodesign Directive concerning electronic products, the use of adhesives is wrongly and unjustifiably subjected to considerable criticism in the context of recycling and the circular economy. The Executive Board regards this as a clear signal that the industry would be very well advised to significantly expand its strategy for communicating with the EU and also with the technical departments of German ministries, other public bodies and associated institutions.

The fact that the German Adhesives Association and its members actively participate in important global conferences underlines the outstanding international position of the German adhesives industry. This applies in particular to the International Adhesive & Sealant Conference that takes place every four years. Following the international conference in Paris in 2012, the German Adhesives Association supported the 2016 International Adhesive & Sealant Conference in Tokyo with a number of interesting presentations. The association has already prepared and registered a block of lectures on the subject of sustainability and the circular economy for the International Adhesive & Sealant Conference which is taking place in the spring of 2020 in Chicago in the USA. Germany's adhesives industry and its association are therefore actively taking the opportunity to highlight to an international audience and to document the techno-logical leadership and the skills profile of the industry in the field of sustainability.

During the Executive Board's regular interactions with adhesives associations in the United States and Asia, it is becoming increasingly evident that the German adhesives industry is not only considered to be a global technological leader but is also regarded with respect for its competence in "Responsible Care®" and sustainable development. In consultation with its system partners, the German adhesives industry began several years ago to develop practical solutions to provide adequate protection for the environment and consumers and to ensure workplace safety, before the relevant legislation was introduced, and went on to successfully implement these solutions in line with the strategic guidelines issued by the Executive Board. By separating solvent consumption from the production of adhesives, successfully establishing the EMICODE® and GISCODE systems and introducing sample EPDs, the German adhesives industry has fulfilled its responsibility to the environment and its customers throughout the entire value chain and is therefore playing a leading role on a global scale. With voluntary initiatives to remove solvents from parquet adhesives and phthalates from paper/packaging adhesives and with its information series "Klebstoffe im lebensmittelnahen Bereich" (Adhesives in the food industry), the German Adhesives Association has repeatedly set new standards in terms of health and safety at work and environmental and consumer protection.

The Executive Board of IVK believes that maintaining a balance in the relationship between technological leadership and social competence is the key to the successful and credible positioning of the industry. This ensures that the association always has access to reliable and important information, for example relating to the future direction of draft legislation, and that members of the association will have the opportunity in the future to open up new markets in Europe and worldwide with products that comply with the relevant requirements on workplace safety and environmental and consumer protection.

The members of the Executive Board view the constantly growing number of participants at the various events organised by the German Adhesives Association and the nine new member companies which have joined the association during the past two years as a clear indication that the association is well positioned, provides a valuable and very practical service for the benefit of its members and is therefore highly attractive to the adhesives industry as a whole.

Technical Board (TA)

In its regular meetings in 2017 and 2018, the Technical Board focused on the specialist work of the technical committees, the subcommittees and the ad-hoc committees, discussed this in detail and developed scenarios for the entire adhesives industry. Furthermore, the Technical Board held in-depth discussions on a number of interdisciplinary topics and issues which were then proactively influenced by the association's corresponding activities.

A major priority of the Technical Board's work has been the subject of ecodesign. Draft documents from the EU Commission, which focused increasingly on reusability, recyclability and repairability, included requirements for disassembly and the publication of instructions etc., but also proposed far-reaching bans on glued joints ("no firm gluing" and "no other gluing than through the use of double-sided adhesive tapes").

In its position paper and in a series of meetings with the German Federal Ministry for Economic Affairs and Energy (BMWi), the Federal Ministry for the Environment (BMU), the Federal Institute for Materials Research and Testing (BAM) and the Environment Agency (UBA), IVK has successfully made it clear that:
- "Bonding" does not prevent a product from being repaired or recycled (end of life). ("Debonding" must be one of the requirements placed on product manufacturers and must be taken into consideration during the product design process, in collaboration with the adhesives supplier.)
- Only a formulation which is technology-neutral (the specification of objectives instead of joining methods) will allow innovations to be developed in future.

After the EU Commission had requested the drafting of the standardisation mandate M/543 "Ecodesign requirements on material efficiency aspects for energy-related products", CEN launched nine new standards projects in this area. This has given rise to a total of four important areas of action for IVK:

- **Ecodesign:** The adoption of the current implementing regulations for displays, servers and computers and for other energy-related products (Directive 2009/125/EC) with the support of IVK and its members.
- **Circular economy package/EU outline legislation on the circular economy:** Involvement of IVK (monitoring/contributing where this is sensible and possible).
- **CEN: standardisation mandate M/543** "Ecodesign requirements on material efficiency aspects for energy-related products" (with nine new standards projects): Involvement of IVK (monitoring/contributing), improved strategies in the light of the very large amount of work required.
- **Communication strategies:** These include the strategic involvement of the relevant technical departments of ministries and public authorities and a change of image: Bonding is not an obstacle to the circular economy, but instead a part of the solution.

The IVK position paper, which is also available in English, was submitted to FEICA, which used it as the basis for a FEICA position paper and brought this to the attention of the EU Commission. At the consultation forum held by the EU Commission on the implementing regulation for displays on 6 July 2017 in Brussels, FEICA presented the position paper and received a positive response. During the forum, the EU Commission withdrew its call for a ban on glued joints and liquid adhesives and is currently revising the implementing regulation.

In addition, IVK has succeeded, with the help of the German Chemical Industry Association (VCI), in incorporating into the current position paper of the Federation of German Industries (BDI) entitled "EU-Ökodesign-Richtlinie: Kriterien der Ressourceneffizienz müssen marktgerecht und widerspruchsfrei sein!" (EU Ecodesign Directive: Criteria for resource efficiency must be market-driven and consistent!) (date: October 2017) the requirements that we regard as being most important:

- Technology neutrality
- Consideration of the material efficiency requirements during the design phase of products (and corresponding transition periods).

The key messages that the adhesives industry has for politicians are therefore as follows:

- **Bonding is not an obstacle to the circular economy, but instead a part of the solution.**
- The use of liquid adhesives or adhesive tapes does not prevent products from being repaired or recycled. If customers require a product to be repairable or recyclable, this can be taken into consideration and achieved during the design of the glued joint.
- The mandatory use of specific technologies constitutes a technological blockade and presents an obstacle to future innovations and the development of more efficient products.
- **Ecodesign requirements** must always be **technology-neutral.** Instead of specifying certain technologies, they must **identify objectives** (see also the BDI position paper "EU Ecodesign Directive: Criteria for resource efficiency must be market-driven and consistent!", principle 4).
- Requirements for material efficiency must generally be incorporated into the **design phase of products.** As suitable solutions cannot usually be integrated into existing designs, a

transition period is needed which will allow manufacturers to develop and implement appropriate solutions for each product in cooperation with their supply chains (see also the BDI position paper "EU Ecodesign Directive: Criteria for resource efficiency must be market-driven and consistent!", principle 10).

In summary, the proposals submitted by the associations met with a positive response. The EU Commission is now seriously working on finding a technology-neutral wording. This is clear from the EU Commission's answer dated 18 September 2017 to the critical comments from South Korea as part of the WTO notification on the "Implementation regulation for displays – G/TBT/N/EU/433". In its response, the Commission wrote: "…4.1. Welding and gluing: the EU will take into the utmost consideration the comments received from different stakeholders not to refer to specific technologies for welding, soldering, brazing or gluing components to be removed before the treatment as provided for by the WEEE Directive (…)." In addition, all the German ministries and agencies (BMWi, BAM, BMU, UBA…) have responded positively and with understanding to the criticism from the adhesives industry.

Another important project is the development of **average EPDs** (environmental product declarations) for the building industry. This work was necessary because since 2011 the EU Construction Products Regulation has specified that "sustainable" products must be used and that the sustainability of products must be substantiated by EPDs. The EPDs, which are required by law, include detailed analyses and documentation, for example with regard to CO_2 emissions during production, energy consumption, resource depletion, ecological assessments, life cycle assessments etc. The preparation of EPDs involves considerable work and substantial costs, which means that an industry-wide approach is the ideal solution. The German EPDs have been completed and offered to other European countries by FEICA.

EPDs have to be revised every five years, which means that the German versions need to be updated. The FEICA EPDs will expire between September 2020 (PU) and August 2021 (DIS). If the German EPDs and the FEICA EPDs are amalgamated, Germany will benefit from the longer lifetime of the FEICA versions. In addition, only one system will have to be revised and the costs can be shared.

The initial analysis of the revised version of EN 15804 "Sustainability of construction works. Environmental product declarations. Core rules for the product category of construction products" produced the following results:
- As the FEICA model EPDs are based on weight units, they are not affected by the change in the rules for function units.
- The FEICA model EPDs meet the requirements of the special rule, which means that the use phase (module C) and the end-of-life phase (module D) do not need to be taken into consideration. Transport on site and installation (A4 and A5) are already included in the existing EPDs.
- Environmental impact indicators: The FEICA model EPDs are based on the following environmental impact indicators: GWP (global warming potential), ADPF (abiotic depletion potential fossil fuels) and POCP (photochemical ozone creation potential). These form the basis for the evaluation of the individual substances. This pragmatic approach must be defended if necessary, particularly as the revision of EN 15804 will lead to the introduction of toxic criteria.

The Technical Board took a highly critical view of the "ProScale" project which was developed by the industry for use with the PEF (product environmental footprint) and EPDs, as a method of carrying out comparative assessments of human exposure (human toxicological data) throughout the value chain. ProScale records all direct exposures. The tool is intended for use in business-to-business and business-to-customer communications and makes use of existing data (generally from REACH). From a political perspective, PEF represents a methodological vacuum. The industry has used ProScale to fill this vacuum in order to be able to report back to the politicians.

ProScale covers the entire life cycle of a product and adds together the different possibilities of exposure throughout the value chain. The Technical Board questioned in particular whether adding together exposures that are regarded in toxicological terms as safe (on the basis of REACH) is permissible, whether this calls the risk-based approach of REACH into question and to what extent the exclusion of indirect exposures (for example, discharging substances via the roof of a building during production) casts doubt on the usefulness of the method. The Technical Board will monitor the ongoing development of ProScale and make its criticisms clear, in particular with regard to the possible consequences for the adhesives industry.

Another focal point of the Technical Board's work over the past two years included the activities relating to the European chemicals legislation REACH (Registration, Evaluation, Authorisation and Restriction of Chemicals), set out in the REACH Regulation (EU) No 1907/2006. France has submitted a CLH report with a proposal for the harmonised classification and labelling of titanium dioxide as "presumed to have carcinogenic potential for humans" (category 1B)/"may cause cancer by inhalation" (H350i). The proposal is in the decision-making process on a European level, with the involvement of the member states. Because of the widespread use of titanium dioxide and the serious effects that it can have, the discussion about this substance has a fundamental significance for the presentation of the consequences of a harmonised classification. The stated effect is not specific to titanium dioxide, but is characteristic of several different types of dust, regardless of the basic material. The German Chemical Industry Association (VCI) and the other industry associations that are affected have submitted the statement on titanium dioxide and the effects of a harmonised classification to ECHA for public consultation. The subject is currently being discussed in detail by CARACAL (Competent Authorities for REACH and CLP) and efforts are being made to reach a practical solution.

Because isocyanates can cause sensitisation of the respiratory tract, the Federal Institute for Occupational Safety and Health (BAuA) has carried out a risk management option analysis (RMOA) of diisocyanates as part of the REACH evaluation. The goal is to reduce the number of occupational asthma cases caused by the use of isocyanates. The industry supports the BAuA's proposal to achieve this goal by means of a restriction (Annex XII, REACH). The restriction is not intended as a general ban on the chemicals. Instead mandatory, verifiable measures will be introduced which will require protective action to be taken and training on the safe use of diisocyanates to be provided. It must be possible to continue using the substances in the case of a total content below a concentration limit of 0.1 percent. Above this concentration limit, objective criteria must be applied to demonstrate that the use involves only a minimal risk, for example because of the safe design of the product. Otherwise training courses must be held. The manufacturers have been offering voluntary courses of this kind for diisocyanate users for several years. The restriction will make this training mandatory and expand its scope. At the same time, users will be obliged to comply with specific regulations concerning the handling of diisocyanates and to provide proof of this. The consequence of a failure to take part in the training programmes would be an automatic ban on using the substances. The restriction dossier also describes a test to determine whether the training measures lead to a reduction in the annual number of sensitisation cases. The goal is a figure of less than 1/500. However, the existing database is not sufficiently robust to allow comparisons to be made. A cohort study to be carried out in Germany by the IPA (Research Institute for Prevention and Occupational Medicine of the German Social Accident Insurance at the Ruhr University Bochum) is intended to create a reliable database. First of all a feasibility study will be completed. The cost of around € 400,000 will be borne by the employers' liability insurance associations and the industry.

Following a long discussion period, on 15 June 2016 the European Commission published proposals for criteria to identify endocrine disruptors as part of the Plant Protection Products Regulation and the Biocidal Products Regulation. The proposed criteria are based on the definition of endocrine disruptors produced by the World Health Organisation (WHO) and the significance of the scientific data (the weight of the evidence). However, the potency of the substances, which in the industry's opinion is a crucial factor in regulatory decision-making, does not form part of the criteria.

In their current form, the criteria cannot be used to distinguish between materials that can be used safely and materials that require regulation. In the industry's view, the future criteria should only cover substances that have a harmful effect on people or the environment in small quantities or doses and therefore require statutory regulation. In order to decide whether a substance is harmful to the endocrine system, several factors need to be taken into account, including the potency of the substance, the seriousness of the harmful effects on an intact organism, the reversibility of the negative effect and the weight of the evidence. Certain member states, for example France, Denmark and Sweden, and some environmental organisations have complained about the fact that the Commission is not planning to introduce additional categories of suspicion (for example, "suspected endocrine disruptor"). In addition, the proposal to investigate "negligible risks" in future as part of the decision-making process was subjected to criticism. In this case, the general opinion is that the Commission is exceeding its authority. The member states are currently discussing the Commission's proposals. We do not yet know what conclusions the Commission will draw from the public consultation and the opinions of the member states or whether changes will be made to the proposals.

There has been and still is a need for action with regard to the European Biocidal Products Regulation (EU Regulation No 528/2012). As part of its evaluation of active biocidal substances that are already in use, the EU Commission is issuing approvals for in-can preservatives in the form of implementing regulations. In particular in the case of substances with sensitising properties, the annex of the implementing regulations for product type 6 (in-can preservatives) generally contains a clause that makes mandatory the additional requirements for labelling specified in Art. 58(3) of the Biocidal Products Regulation. The decisive factor in determining the deadline for implementing the labelling requirements is the date of approval of the substance. Only then do the special provisions of the implementing regulation come into force. IVK has informed its members in a number of newsletters about the approval of various in-can preservatives, the resulting changes in labelling and the accompanying deadlines. As an additional service, it has made available a table on the IVK intranet which gives an up-to-date overview of the in-can preservatives that have been newly approved by the EU Commission. The table indicates whether the implementing regulation for the active substance used as an in-can preservative in adhesives contains new labelling requirements in accordance with Art. 58(3) of the Biocidal Products Regulation and when these must be implemented.

The **Regulation on harmonised information relating to emergency health response** was published as Annex VII of the CLP Regulation in the Official Journal of the European Union on 23 March 2017. According to this document, all mixtures placed on the market and classified as hazardous must be reported. (The only exception is that there is no reporting requirement for development and testing). In the case of mixtures used in industry, the information from the safety data sheet as specified in REACH Annex II is sufficient, provided that there is a 24/7 information desk available.

Very detailed reporting obligations apply to products for end consumers and professional users. A UFI (unique formulation identifier) is mandatory for all products. In the case of products not intended for consumers, it can simply be included in the safety data sheet. ECHA has made available a tool that can be used free of charge to generate the UFI. Any change in the formulation that results in a new registration will also require a new UFI. During the registration process, all the known components of the mixture (> 0.1 % with critical hazard classifications and > 1 % without critical hazard classifications) must be specified, with the exact chemical name and CAS number. Consumer products must be reported to the poison centres by 1 January 2020 (this deadline will probably be extended to 1 January 2021), products in professional use by 1 January 2021 and products used in industry by 1 January 2024. In the case of mixtures that have already been reported under existing regulations in the member states, the poison information report does not need to be updated until 1 January 2025, provided that no new safety information is available.

IVK members obtain information relating to REACH/CLP and the Biocidal Products Regulation via the VCI's REACH portal as well as in the newsletters published by IVK to keep its members informed about adhesive-related issues and the seminars it runs in cooperation with other organisations.

On the basis of an initiative by the German Federal Ministry for the Environment (BMU) to reduce VOCs due to the risk of summer smog, VCI was asked, together with other industrial associations, to explore the possibilities for reducing the use of VOCs. Driven by concern for the environment, this initiative has led adhesive manufacturers to reduce their solvent consumption significantly over the past few years, as an evaluation of the IVK solvent statistics shows. The Technical Board has addressed this topic in depth and was able to demonstrate with the aid of its solvent consumption survey that the objective of reducing VOCs by 70 % by 2007 (based on the consumption figures in 1988) was in fact reached much earlier. The solvent statistics are compiled every two years and used as the basis for discussions with German and European agencies and institutions, for example when introducing new legislative proposals. The statistics are a very important instrument for communicating with agencies and authorities and are indispensable for documenting the environmental awareness of Germany's adhesives industry.

At the end of February 2016, the DIN standard 2304 "Adhesive bonding technology – Quality requirements for adhesive bonding processes – Part 1: Adhesive bonding process chain", which IVK played a significant role in developing, was published with a date of March 2016. This represents another important means of safeguarding the quality of the bonding process chain. The new standard specifies the requirements for the high-quality professional production of constructional/structural/load-bearing bonded joints in pre-defined safety classes. The definition of the safety classes is based on the possible risks of injury or death in the event of the mechanical failure of a bonded joint. The standard does not cover the requirements for bonded joints with regard to suitability for use with food, fire prevention or compliance with emissions regulations or health and safety provisions. In addition to the documentation of the bonding process, the standard also requires knowledge of and training in the use of adhesives. This means that the training in adhesives on the basis of the safety class of the bond, which in the past was voluntary, has now become a mandatory component of controlled bonding processes. The tasks and responsibilities of bonding supervisors are specified in DVS guideline 3311. Companies that manufacture products in future in accordance with this standard must also document the fact that they are using the latest bonding technology. The Technical Board has commissioned a presentation/lecture on DIN 2304. It is available in four different and comprehensive versions and can be used by adhesive manufacturers to explain DIN 2304 to their customers and to provide adhesive users with adequate information about the implementation of DIN 2304 at conferences, seminars and other events.

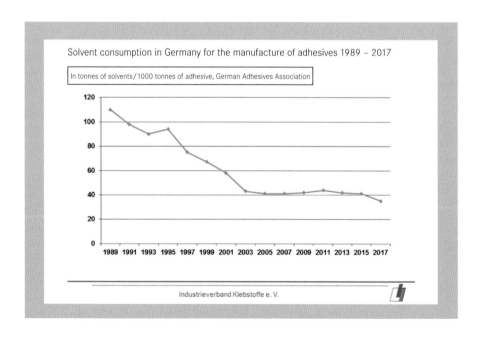

Solvent consumption in Germany for the manufacture of adhesives 1989 – 2017

In tonnes of solvents/1000 tonnes of adhesive, German Adhesives Association

Industrieverband Klebstoffe e. V.

The Technical Board has also focused in detail on the subject of adhesives training. It strongly recommends taking steps to extend the training of adhesives users and to promote the training throughout Europe with the aim of increasing the number of qualified employees on the market. In this context, the Technical Board supports a project by Fraunhofer IFAM to establish a certified training programme for adhesives engineers on an international level. The Master of Engineering (MEng) course in bonding as a joining technology is still in the pilot phase. The organisation responsible for the course is Steinbeis/SMT and it is being run by Steinbeis/SMT and Fraunhofer IFAM.

Since the appearance in 2002 of the first series of slides entitled "Bonding", which was produced in collaboration with the German Chemical Industry Fund, the subject of adhesives has fortunately been incorporated into the science curriculum and, in particular, the chemistry curriculum for secondary schools. As a result of the high level of demand, the comprehensive teaching materials have been completely revised and brought fully up-to-date in terms of both content and teaching methods, once again in collaboration with the German Chemical Industry Fund. The new teaching materials on the subject of "Die Kunst des Klebens" (The Art of Bonding) have been distributed to almost 17,000 technology teachers throughout Germany. Outside the school classroom, these materials are an excellent means of supplementing the content of company training courses and presentations.

In 2016 the Standing Conference of the Ministers of Education and Cultural Affairs of the German states adopted the strategy paper "Bildung in der digitalen Welt" (Education in the digital world), which sets highly ambitious objectives. The aim is to enable children and young people "to live an independent and responsible life in a digital world". In order to meet these objectives, digital infographics have been included in the information series. The IVK teaching material can be linked with the other information provided by the association (job profiles for the adhesives industry, the promotional video, the adhesives guidelines). This gives education specialists a tool that meets the new requirements for digital education. The teaching material will be made officially available for the school year 2018/2019. Alongside the digitisation process, the adhesives educational materials have been revised and expanded by the University of Frankfurt specifically for the purposes of teacher training. These materials will also be included in the digital package.

The adhesives industry is one of the first sectors of the chemical industry to offer teachers a digital package of this kind. This increases the likelihood that the subject of bonding and adhesives will be included in the school curriculum.

The Technical Board has revised the entries for "Kleben" (bonding) and "Klebstoff" (adhesives) in the German version of Wikipedia.

Other topics discussed by the Technical Board included:
• Classification and labelling issues
• Health and safety at work, environmental and consumer protection issues
• Standards activities
• Monitoring various EU Commission projects in Germany

Technical Committee
Building Adhesives (TKB)

Introduction

The Technical Committee Building Adhesives (TKB) of the German Adhesives Association (IVK) represents the interests of manufacturers of building adhesives and dry mortar systems who are members of IVK. The committee liaises with public authorities, trade bodies, employers' liability insurance associations, other industry organisations and standardisation committees.

Its aims are to establish technical standards, to influence the provisions of legislation on chemicals, to participate in developing legislation, to promote technical progress while safeguarding users and the environment, to provide technical support and information to customers and the building trade, and to promote the use of building adhesives and mortar systems by providing objective technical information.

Overview of topics

TKB's activities can be broken down into several categories:
- Technical topics relating to building adhesives/flooring installation products and their applications.
- Standards for flooring and parquet adhesives, levelling compounds, tile adhesives, binders for floor screeds, primers and special products.
- Technical information events for the flooring and parquet laying trade and other related trades.
- TKB publications about current topics relating to application methods and building legislation and about standardisation and environmental and user protection issues.
- Building legislation issues, including approvals in Germany and Europe.
- Topics relating to chemicals legislation, such as German, European and international regulations on labelling hazardous materials and REACH.
- Health and safety at work and environmental and consumer protection.

The committee's work

Mortar systems

On the national and international standards bodies ISO/TC 189/WG3, CEN/TC 67/WG 3, CEN TC 303, WG 2, NA 062-10-01 AA and NA 005-09-75 AA, representatives of TKB help to define the standards for flooring/parquet adhesives, tile adhesives, levelling compounds, sealing compounds and floor screeds.

The requirements standard (EN 12004-1) for tile adhesives was published in April 2017. However, it has not yet been published in the Official Journal of the EU, which means that annex ZA of the old standard still applies for CE marking.

The situation for sealing compounds is similar. The revised DIN EN 14891 standard was published by the Beuth Verlag in 2017, but it has not yet been published in the Official Journal of the EU, so in this case too annex ZA of the old standard still applies for CE marking. The draft versions of the revised test principles for sealing compounds have been published on the website of the DIBt

(Deutsches Institut für Bautechnik). However, because of the formal processes required, these will not come into force until at least 2022.

The sealant standards in the series DIN 18531 to DIN 18535 were published in July 2017. The DIN 18195 standard will continue to be available purely for definition purposes. At the end of 2018 the Beuth Verlag published a commentary on these standards which includes users' experiences of the standards to date. The standards themselves will only be revised as part of the five-year revision cycle.

Product and application technology
TKB continued its comprehensive investigations into the subject of screed drying, the readiness of screeds for the installation of floor coverings and subfloor moisture measurement methods. It commissioned Hamburg University of Technology to carry out a study to determine the scanning isotherms of mineral screeds and to investigate the influence of temperature on the corresponding relative ambient humidity. The results were presented at the TKB conference in 2018 and then published as TKB report 4. In 2017 a comprehensive ring test was carried out as a practical evaluation of the KRL method in collaboration with several well-known industry experts. This was continued in 2018 and CM and KRL moisture levels were measured in parallel on building sites. A total of 260 measurements were taken with more than 20,000 data points. These results were presented at the TKB conference in 2019 and published as TKB report 5. The existing TKB guideline figures were converted into thresholds for the readiness of mineral screeds for the installation of floor coverings. The process of taking measurements using the KRL method is described in TKB technical information sheet 18 which was published in August 2018.

Standards
In cooperation with the IVK standards competence centre, TKB has helped to define international, European and German standards as part of the ISO/TC 61/SC 11/WG 5, CEN/TC 193/WG 4, NA 062-04-54 AA and NA 005-09-75 AA standards bodies. The work on extending mandate M/127 to convert EN 14293 (Parquet adhesives) and EN 14259 (Adhesives for floor coverings) into harmonised European standards is still underway. The scope of EN 14259 is to be extended to include dry adhesives and fixings. However, TKB has stopped any further work on these standards. Instead it has increased its efforts to bring the content of these standards up to the ISO level (ISO 17178 and ISO 22636). As a supplement, a further ISO standard concerning the determination of the emission behaviour will be introduced on the basis of the GEV testing method.

In the case of levelling compounds, TKB representatives are involved via NA 005-09-75 AA and CEN/TC 303 in revising EN 13813. The German draft version of the revised standard was published in March 2018. This specifies that an emissions analysis is only needed for resin screeds. There are no requirements of this kind for mineral screeds. The requirements standard EN 13892-9 relating to the shrinkage of screeds has been integrated into the EN 13813 standard. This standard has been approved by the CEN technical committees. Following a number of objections by the EU Commission, publication in the Official Journal of the EU has been delayed. Cooperation between CEN/TC 193, which drew up the test standards for levelling compounds, and CEN/TC 303 is guaranteed by the membership of the technical bodies (liaison). A new part (part 8) concerning decorative screeds is planned for the DIN 18560 series of standards. This will also cover decorative levelling compounds. The revision of part 2 of the standards series (Floor screeds on insulation layers) began in 2018.

The main test standards for the floor covering standards (EN 1372, EN 1373, EN 1902 and EN 1903) have been transferred to the ISO level. In preparation for the work on standardising the KRL method, another EMPIR funding application was submitted. The accompanying standardisation project has been accepted by CEN.

Events and publications

The 34th TKB conference in 2018 and the 35th in 2019 covered topics from the fields of product technology, building legislation, health and safety at work and the environment. In 2018 presentations were given on the subject of TKB technical information sheet 14 "Classifying cement screeds", on laying decorative floor coverings in bathrooms, on damage to floor coverings caused by faults in the structure of the floor, on the technical requirement for system recommendations and on the functional role of packaging. In addition, the latest results of the investigations into the KRL method at Hamburg University of Technology and the results of the ring test involving industry experts were presented. A panel discussion was held for the first time at a TKB conference. Representatives of the building trades discussed the KRL method and the audience had the opportunity to ask questions. In 2019, there were presentations on the effectiveness of screed additives, on subfloors in renovation work, on thermally activated component systems, on heat-sealing joints in flexible floor coverings, on the development of floor laying materials over time and on BIM (building information modelling). Once again a panel discussion took place, this time on the subject of laying floor coverings on old subfloors.

The TKB industry meetings which were held each autumn were attended by the editors of the main industry magazines and covered current technical and regulatory issues. Other guests were members of important industry associations, experts and representatives of the floor covering industry. The subjects under discussion included the commentary on DIN 18365, the approval situation for floor laying materials, cooperation on the screed standards process, definitions of terms relating to fixings and anti-slip products, system recommendations, smoothness tolerances for levelled subfloors, the KRL method, laying floorings on stairs without the use of solvents and the importance of sustainability and the Blue Angel certificate for floor laying materials.

The process of revising the TKB technical information sheets has largely been completed. The revised version of TKB technical information sheet 9 "Floor levelling compounds" was published in June 2019 and TKB technical information sheet 6 "Toothed scrapers" was published in March 2019. The contents of technical information sheet 6 are currently being incorporated into the international ISO standard. TKB technical information sheet 17 "Effects of the indoor climate on floor coverings and floor laying materials during laying and use" was also revised. Because of it describes the use of floor coverings, it was revised in collaboration with the German Association of Parquet and Flooring Installation Companies (BVPF), which drew up its own shorter version in the form of a reference sheet for users. This is the first TKB technical information sheet to be available only as a download. The TKB technical information sheet entitled "The KRL method for measuring moisture levels to determine the readiness of mineral screeds for the installation of floor coverings" was published in August 2018. The work on the KRL method was described in TKB report 4 "Sorption isotherms and the interpretation of KRL measurements" and TKB report 5 "A ring test of measuring moisture with the KRL method".

Statements from TKB on current subjects, including new and revised TKB technical information sheets, were published in the industry press and on the IVK website under the heading "TKB informiert..." (Information from TKB...). Two information sheets on the readiness of screeds for the installation of floor coverings and on special structures were published in collaboration with the BVPF.

Cooperation with other associations/institutions
TKB consults with a variety of associations and organisations on technical and regulatory issues. An informal contact was maintained with the German Federal Association of Screeds and Floor Coverings (BEB) via some TKB members who are also supporting members of this organisation.

An informal contact was maintained with the German Association of Manufacturers of Flexible Floor Coverings (FEB) via some manufacturers of floor laying materials who are also supporting members of this organisation.

The collaboration with the German Association of Parquet and Flooring Installation Companies (BVPF, formerly ZVPF), which is the most important body in this sector, proved to be very useful, largely due to the involvement of TKB in the BVPF expert advisory committee. The technical and regulatory issues that affect the trade and the installation materials it uses will now be covered regularly in the BVPF expert conference, as well as in the TKB conference. In a joint press release aimed at a technical audience on the increase in the moisture thresholds of heated calcium sulphate screeds, TKB and the BVPF explained how technical discussions should be used to support the financial interests of individual groups, in this case parquet and flooring installation companies. The KRL ring test to determine moisture levels in parallel using the CM and the KRL method was largely supported by the BVPF experts. The two organisations jointly published TKB technical information sheet 17 "Indoor climate" and the two information sheets on the readiness of screeds for the installation of floor coverings and on special structures.
The German Federal Association of Underfloor Heating (Bundesverband Flächenheizung) published a revised document on coordinating interfaces for underfloor heating in existing buildings in June 2018 with IVK as a supporting organisation.

On a European level and in particular as part of the FEICA Working Group Construction, TKB representatives have contributed to the implementation of the Construction Products Regulation, supported the introduction of European standards for floor laying materials and promoted the use of German sample environmental product declarations (EPDs) in a European context. TKB's involvement in FEICA ensures that it receives information about planned European regulatory measures at an early stage, such as the revision of the Construction Products Regulation, the obligation to provide information to national poison centres, the restrictions on the use of products containing diisocyanates and the circular economy. In the Smart CE Labelling project, which will in future make information about CE marking and additional product data available via a QR code on the product, a standardised data format was defined on the basis of a CEN workshop agreement.

Issues relating to building legislation

The in-depth dialogue with the DIBt about the regulation of sealing compounds is ongoing. Work on revising EN 14293 and EN 14259 has been stopped to ensure that the regulation of adhesives for floor coverings and parquet will continue to be part of the general technical approval process. In October 2015, the European Court of Justice (ECJ) passed its judgement C-100/13, which declared the additional approval requirement for CE-marked products to be invalid. As a consequence of this, the Deutsche Institut für Bautechnik (DIBt) published the sample administrative regulation on Technical Construction Conditions in August 2017. This has subsequently been incorporated into the building regulations of all the German federal states. It specifies that CE marked products must not be required to have an additional mark of conformity. As adhesives do not need CE marks and levelling compounds are regulated by EN 13813, which does not apply any emission requirements to mineral products, the adhesives industry is almost entirely unaffected by this new regulation.

TKB is playing an active and critical role in support of the efforts of the German Environment Agency (UBA) to introduce odour testing for building products. In addition to TKB's involvement in the sensors working group of the Committee for Health-Related Evaluation of Building Products (AgBB-AG Sensorik), a study was carried out by the Fraunhofer Institute for Wood Research WKI to determine whether a valid method for identifying odours was already available. As things currently stand, no requirements relating to the odour of building products will be incorporated into the DIBt regulations. However, during the changes to the criteria for awarding Blue Angel certification made at the end of 2018, a sensory test was introduced for products that are claimed to be almost odour-free.

Hazardous materials/health and safety at work

The GISCODE group for epoxy resins has been revised jointly by BG Bau (the German employers' liability insurance organisation for the construction industry), TKB and Deutscher Bauchemie (the German association for manufacturers of construction chemicals). In future there will be twelve RE groups and only four of these will be relevant for floor laying materials. Following the introduction of the new group in mid 2018, a two-year transition period began during which the product labelling must be changed. Two additional GISCODE groups will be created for calcium sulphate levelling compounds that require labelling: CP 2 – alkaline and CP 3 – highly alkaline.

Together with BG Bau, TKB will continue carrying out investigations into the methanol emissions resulting from the use of silane-based primers to determine whether compliance with occupational exposure limits is needed, in a similar way to GISCODE RS 10 products, or whether another GISCODE group should be introduced. Because of the plan to halve the threshold limit value (TLV) for methanol, new measurements of methanol exposure from silane-based adhesives on building sites are needed to confirm that the thresholds can be met reliably.

A ranking system for epoxy resin products is also being developed in collaboration with BG Bau (EIS – epoxy information system). Subsequently the aim is to assess the risk potential of the different products on the basis of the substances they contain using a computer model. The accompanying computer program is available on the website of the Federal Institute for Occupational Safety and Health (BAuA).

On a European level, a decision has been made to tighten up the labelling of products containing methylisothiazolone. This measure will take effect in 2020. Manufacturers will need to carry out extensive reformulation work in order to be able to continue supplying dispersion products without special labelling.

The environment/consumer protection/sustainability

FEICA was responsible for transferring the German sample EPDs (environmental product declarations) to a European level and for introducing them throughout Europe. The validity of EPDs is limited to only five years, which means that the German sample EPDs would have needed to undergo approval again soon. In order to keep costs to a minimum, in future the European sample EPDs will be used in Germany and the German sample EPDs will be allowed to expire. The guidelines for awarding Blue Angel certification (RAL-UZ 113) have been revised and made much more stringent from the perspective of manufacturers of floor laying materials. The emission thresholds have been lowered, more extensive restrictions have been imposed on plasticisers and the option of an odour test has been introduced. In addition, strict rules have been applied on the disclosure of formulations.

Technical Committee
DIY and Consumer Adhesives (TKHHB)

The Working Group and Technical Committee DIY and Consumer Adhesives (AGHHB/TKHHB) regularly hold joint meetings and monitor a variety of legislative activities on a European and national level. Their main interest is in regulations and topics concerning adhesives in small packages which are intended for private end users. Important topics in this area include:

- Identification and packaging of adhesives and the obligation to provide information (CLP Regulation)
- Requirements in certain areas where adhesives are used (Appliance and Product Safety Act/Toy Safety Ordinance/Medical Products Act)
- Other standards-related/legislative restrictions and requirements regarding adhesives

Identification and packaging of adhesives and the obligation to provide information (CLP Regulation)

According to Article 4 of the CLP Regulation, the manufacturer or importer must:
- classify substances and preparations prior to marketing them
- package them according to the classification and
- label them.

The Technical Rules for Hazardous Substances (TRGS 200) summarise the relevant regulations:
- special labelling regulations for substances and preparations that are available to the general public, 6.7; 10.2

- simplification of labelling requirements and exceptions, 7.1
- implementing labelling regulations, 9

On 23 March 2017 Regulation (EU) 2017/542 of 22 March 2017 amending the CLP Regulation (Regulation (EC) No 1272/2008 on classification, labelling and packaging of substances and mixtures) by adding an Annex on harmonised information relating to emergency health response was published. This Regulation came into force on the twentieth day after its publication in the Official Journal of the European Union.

The first deadline is **1 January 2020.** It applies to the submission of information about hazardous mixtures that are placed on the market for consumer use. Further staggered deadlines apply to the requirements for submitting information about mixtures that are placed on the market for professional and industrial use. These are **1 January 2021 and 2024** (submissions made under previous regulations will remain valid until **1 January 2025** unless changes are needed). Paint and coating manufacturers will have to modify their IT systems and internal processes in order to be able to create the files (in XML format), which have to be submitted electronically, before the relevant deadline. These regulations differ from those for labelling or for safety data sheets and therefore additional processes will be needed. In addition, manufacturers will have to begin including a UFI code (unique formula identifier) on the labels of all the products that they place on the market and that are classified as hazardous under the terms of the CLP Regulation.

UFIs are created from a VAT number and a numeric formula code. A free tool is available on the ECHA website for creating UFIs. It will also be possible to integrate the UFI creation process into companies' internal IT systems. The UFI code will probably change more frequently than the product labelling. As a result, many companies will need to add the code to the label in the packaging machine which could lead to additional investments being required.

The objective is to harmonise on an EU level the information for the emergency health response that the bodies appointed by the member states under Article 45 of the CLP Regulation receive from the importers and downstream users and to specify a format for submitting the information.

The new Annex VIII of the CLP Regulation on harmonised information relating to emergency health response and preventative measures results in new requirements for submitting information, new labelling requirements, the inclusion of new information in the safety data sheet and the creation of formula identifiers etc.

Companies can obtain comprehensive information about Annex VIII of the CLP Regulation from the following page on the European Chemicals Agency (ECHA) website: https://poisoncentres.echa.europa.eu/de/steps-for-industry .

Requirements in certain areas where adhesives are used (Medical Products Act/Appliance and Product Safety Act/Toy Safety Ordinance)

As a general rule, a product may only be sold if it is designed in such a way that the health and safety of users or third parties are not at risk if it is used as intended or in a manner that can be foreseen.

This affects the manufacturers of items that are ready for use. Adhesives are not directly subject to this regulation but are indirectly affected via the end product if the safety (and usability) of an item depends on the suitability of the adhesive.

Adhesives as toys or components in toys for the purpose of the Toy Safety Ordinance/ Appliance and Product Safety Act – Toy Safety Directive – EN 71.

EN 71 was revised on behalf of the EU Commission. It is based on the safety requirements of the Toys Directive 88/378/EEC, which requires toys, in other words products which are intended for children younger than 14, to be safe before they are placed on the market. Compliance with the requirements defined in EN 71 is documented by the CE mark. Tests can be carried out by the manufacturer or by a testing body. Alternatively an EU type approval test can be performed if compliance with EN 71 cannot be established in any other way.

Product information in emergencies (DIN EN 15178), labelling checklist

The purpose of the product information standard DIN EN 15178:2007-11 is to improve the identification of products in the event of emergencies. The letter "i" placed near the bar code on packaging refers to the trade or product names or the number under which the product is registered or officially approved.

The Technical Rules for Hazardous Substances (TRGS 200) "Classification and identification of substances, preparations and products" provides another template for a **checklist for labelling issues,** which has been revised by TKHHB.

New Packaging Act

The AGHHB has also been working on the Packaging Act which came into force in 2018 in place of the previous Packaging Ordinance and which sets ambitious recycling targets. The obligation to take part in a dual system for packaging remains unchanged. This "typically" applies to end consumers or equivalent parties. There are insurmountable obstacles for industry solutions, including the requirement to list all customers. Exceptions are only permitted for existing industry solutions, which involves treating them as bulk materials containing hazardous substances and allowing for separate collection. The establishment of a "central body" as a "packaging register" means that IVK is no longer required to act as a reporting body. Alongside the familiar declaration of completeness, used packaging will also be registered and reported to the central body.

In a purely commercial context, there is no requirement to take part in the system. This makes it possible to use a more cost-effective commercial system to recover the packaging or to return it to the supplier.

Construction legislation

New regulations in the field of construction legislation also affect the working group. National regulations on construction products and on labelling, like those in France and Belgium, and approval regulations issued by DIBt in Germany are important, because in some cases they affect DIY products. A VOC classification system is being considered on a European level that will be incorporated into the declaration of performance (CE) of the products via harmonised standards.

Since 2018 construction contracts in Germany have no longer been covered by the legislation on work contracts, but have their own provisions (sections 650a ff. of the German Civil Code). As a result, builders have additional obligations and rights, for example the specification of a completion date and a 14-day cancellation period. In addition, construction companies must provide comprehensive information about plans, approvals and certifications, for example. If individual services do not correspond with what was promised, the construction company will be responsible, unless a clear statement to the contrary was made in the construction contract. Provisions have also been introduced concerning the amounts of payments by instalment.

Technical Committee
Wood/Furniture Adhesives (TKH)

In recent years, health and safety at work, environmental protection and consumer protection have become increasingly important issues. One example of this is the new guidelines and thresholds for the use of biocidal products. These will have an impact on wood adhesives, which contain biocides as in-can preservatives. The contacts that the Technical Committee on Wood Adhesives (TKH) has with associations and institutes are very important and have been of valuable help over the years.

Over the last few years, TKH meetings have been regularly held on the premises of a variety of bodies, such as the Institute for Wood Technology (IHD) in Dresden and the ift Rosenheim (the Testing Institute for the Evaluation of the Fitness for use of Construction Products).

The TKH expert panel is currently working on a new data sheet on the subject of avoiding delamination in various applications and using different types of adhesive. The first area under consideration will be bonding to solid wood.

Data sheets entitled "No CE marking for wood adhesives" and "Interpreting DIN EN 204/205 during building inspections" have been published recently. The existing series of data sheets covering specific adhesive-related subjects is available to download in German and English from the IVK website.In addition to its internal activities, TKH also takes part in working groups covering a number of different industries, such as the profile wrapping working committee and the initiative group on three-dimensional furniture fronts.

The RAL expert group on profile wrapping decided to update RAL 716 and, for example, to provide more accurate information about the conditions for the peel test. The revision of RAL-GZ 716 was completed in 2019. The 3D Furniture Front Production Quality Guide, which was completed in May 2017, is now available in Chinese. Further language versions, including Turkish and Russian, are also planned.

TKH is contributing its expertise to the revision process for a set of VDI (Association of German Engineers) guidelines "VDI RL 3462-3 Reducing emissions – Wood processing and woodworking; Processing and finishing wood materials".

TKH is regularly represented on the CEN TC 193 SC1 Adhesives for Wood standards committee and its subcommittees. One important development was the publication of the standards DIN EN 204 (Classification of thermoplastic wood adhesives for non-structural applications) and DIN EN 205 (Adhesives. Wood adhesives for non-structural applications. Determination of tensile shear strength of lap joints) as ISO DIS 19209 and ISO DIS 19210.

CEN TC 193 SC1/WG 12 is in the process of revising the standard DIN EN 14257:2006-09 "Adhesives. Wood adhesives. Determination of tensile strength of lap joints at elevated temperature (WATT '91)". TKH is continuing to take part in cooperative tests as part of the development of the standard for classifying wood adhesives for non-structural applications in outdoor use.
It is very important that standards and proposed standards are monitored with regard to their practical relevance.

TKH can only make use of the collective expert knowledge of the industry for the benefit of users with the commitment of the member companies and as a result of their willingness to give their employees the opportunity to work on behalf of the association.

Technical Committee Adhesive Tapes (TKK)

The main role of the Technical Committee Adhesive Tapes (TKK) is to initiate and support publicly funded research projects. These projects are motivated by enquiries from industries that use adhesive tapes and also by the general recognition that, in comparison with liquid adhesives, for example, there are very few research projects on the subject of adhesive tapes with publicly available results. Ensuring that a broader range of research is carried out is an essential requirement for increasing the use of adhesive tapes in more technically challenging applications and is therefore another driving force behind the work of the TKK.

The long-term durability of bonds is particularly important in the automotive industry and also the construction industry. It goes without saying that it must be possible to provide reliable information about the long-term behaviour of bonds created using adhesive tapes. The relevant tests are generally costly and highly time-consuming and, as a result, the development cycles are very long. For this reason, TKK has worked with the Fraunhofer Institute for Manufacturing Technology and Advanced Materials (IFAM) to apply for funding for a research project in this area with the aim of reducing the duration of a 50-day test to just a few days without introducing more radical test conditions. The project was run by IFAM as part of an AiF (German Federation of Industrial Research Associations) research programme. The results showed that physical test methods can highlight changes in PSA (pressure sensitive adhesive) systems caused by thermal or hydrothermal ageing at an early stage, but this does not necessarily make it possible to draw conclusions about the results of a long-term ageing test. One central issue in this context is the applicability of the Arrhenius relationship with regard to the reaction kinetics of PSA systems. It is clear that there is a need for more fundamental research. The requirement has been structured and formulated by TKK, together with IFAM, in order to justify a follow-up to the project described above. The contribution made by TKK included proposals for selecting relevant and representative adhesive tape systems and also suggestions for test scenarios. A project outline prepared by IFAM was discussed by GAK (the joint committee on adhesives technology) and was very well received, in part as a result of the dedicated support given to it by TKK. The project has now started and the first project meeting was held in June 2019. Several member companies of TKK are providing practical support for the project in the form of model systems.

At the same time, TKK has initiated another research project relating specifically to adhesive tapes. This also concerns the requirements of vehicle manufacturers and, in particular, the public transport sector. It involves the fire protection characteristics of adhesive tapes, including flame retardance and smoke properties. As far as the application is concerned, the subject of flame retardance is a very complex one in the context of public transport. Given that the primary function of flame retardance is to allow passengers the opportunity and the time to escape from a burning vehicle, there is a very wide range of requirements profiles, test methods, test standards and test scenarios which are specifically designed for individual use cases and are generally highly complex. This makes it almost impossible to compare the results of tests across this field. As far as the products are concerned, there is an equally broad variety of chemical and physical influences. Some of these reinforce one another and others impede one another in relation to the flame retardance of specific additives, for example. In this case too, there is no one single system, but instead a range of individual empirical solutions. Finally, the design of the bonded joint also plays a decisive role in a component's reaction to fire. Following in-depth discussions with TKK, the Fraunhofer IFAM has prepared the outline of a project on this subject which focuses primarily on the adhesive-tape-related design of components that have to meet flame retardance requirements. The aim of the project is to draw up guideline test scenarios for adhesive tapes and components for use in development projects. The project outline has been recommended for funding by GAK, preparatory work has begun and the project itself is likely to start in 2019.

Another area where the TKK is active relates to industry-specific customer requirements. The committee is currently working on the technical details of a set of general delivery specifications for wiring loom tape for use in vehicles. Adhesive tape manufacturers have identified the need to review some of the points in these specifications. The TKK has set up an ad-hoc group for this purpose which is working on the content and on the communications with the German manufacturers. For these companies, the delivery specifications play a very important role.

In the field of standards, TKK has provided support for a project that has been initiated by the DVS (German Welding Society). It involves drawing up user guidelines for the quality requirements for the use of double-sided pressure sensitive adhesive tape for permanent bonded joints. The DVS 3320-2 guidelines have now been published in both German and English.

TKK is also involved in activities concerning standards issues that relate to measurement processes on a European and an international level.

The different regional methods for measuring adhesive strength, shear resistance and tensile strength have been harmonised in collaboration with the USA and Japan. Afera (the European Adhesive Tape Association) was responsible for submitting the new harmonised methods to the ISO secretariat for use in countries throughout the world. The harmonisation process has now been completed on a European level (CEN) and an international level (ISO 29862, ISO 29863 and ISO 29864).

Attempts are currently being made to take over the ISO methods as EN and DIN methods. CEN has not yet succeeded in achieving five votes in favour of this on the voting panel. Despite a personal letter being written by L. Jacob to all CEN country representatives, only three of them have voted for this proposal.

As has already been reported, new test methods are being developed by the Global Tape Forum (GTF) which will subsequently be published as GTF test methods. The **Shear Adhesion Failure Test Method** is currently being introduced globally as GTF 6001. At the GTF meeting held in Beijing in May 2016, the Global Test Method Committee decided to adopt the Thickness and Width and Length test methods, which had already been harmonised on a worldwide basis, as new GTF test methods 6002 and 6003. A vote will be held on this subject at the GTMC meeting in Munich in June 2019. In addition, JATMA (the Japanese Adhesive Tape Manufacturers Association) has proposed the development of a new test method for measuring adhesive strength on different steel sheet surfaces.

The Pressure Sensitive Tape Council (PSTC) is now developing a test method for dynamic shear which will be used as the basis for a global method. Details of this test will be presented at the GTMC meeting in Munich.

Afera has carried out several ring tests using the new **Loop Tack Test Method.** The results have been good, but are not yet sufficiently accurate. During the most recent ring test, the length of the plateau that the average figure is based on was redefined. The plan is for the new method to produce two results: the plateau and the figure for the separation of the tape from the test plate. The results of this ring test, which look very promising, will be presented at the GTMC meeting in Munich.

The PSTC has drawn up a proposal for harmonising and introducing new GTF test methods. A vote will be held on this at the global meeting in Munich.

The PSTC has developed a statistical method for determining measurement uncertainties when using test methods. The method can be directly taken over by Afera, but this involves a large amount of work, because ring tests with at least eight participants have to be carried out for each test method.

Afera has produced a new version of its Test Methods Manual, which contains the latest adaptations and the GTF test methods.

As part of its activities, TKK is continuing to support Afera. TKK also contributes reports on conferences and events organised by the adhesive tape industry and on new products to the Afera News newsletter.

Technical Committee
Paper and Packaging Adhesives (TKPV)

Adhesives for products intended to come into contact with food

As in previous years, the Technical Committee Paper and Packaging Adhesives continued to focus on adhesives for food contact materials during the 2018/2019 period. It is still the case that adhesives used for producing materials and articles intended to come into contact with food are not specifically covered by EU regulations. Due to the fact that adhesives are part of food contact materials, they are, however, subject to assessment under the provisions of food legislation (Framework Regulation (EU) No 1935/2004). If adhesives used for products that come into contact with food are based on materials which are also used to produce plastics for products that come into contact with food, information from the corresponding EU regulations can be used for a risk assessment based on Article 3 of Regulation (EU) No 1935/2004. In this context, Regulation (EU) No 10/2011 came into force on January 2011 and replaced Directive 2002/72/EC of 6 August 2002. It consolidated all the material lists from the annexes and all the changes and amendments in one regulation. This has been added to and/or corrected approximately twice a year since it came into force.

For substances that are not listed there, the national European regulations, such as the recommendations of Germany's Federal Institute for Risk Assessment (BfR) or the Royal Decree RD 847/2011, still apply. Regulation (EU) No 10/2011 states that adhesives may also be made of substances that are not approved in the EU for the production of plastics.

Regulation (EC) No. 2023/2006 on good manufacturing practices for materials and articles specifies the principles of good manufacturing practices (GMP), as required by Article 3 of Regulation (EU) No 1935/2004 (EU Framework Regulation for Materials and Articles Intended to Come into Contact with Food). This regulation applies to all articles and materials listed in Annex 1, including adhesives. Strictly speaking, the GMP regulation does not apply to raw materials that are used in the corresponding adhesives. Nonetheless, the materials do have to fulfil specifications which enable adhesive manufacturers to work in accordance with GMP. Guidelines entitled "Good Manufacturing Practices" are available to help adhesives manufacturers to comply with the regulation.

The Union Guidance on Regulation (EU) No 10/2011 was published on 28 November 2013 by the services of the Directorate-General for Health and Consumers and is available in English only. It is intended to help with interpreting and implementing questions about declarations of conformity and other conformity activities and with providing information along the supply chain for food contact materials.

In this EU guidance document, adhesives for the manufacture of food contact materials made of plastic or plastic composites are described as "non-plastic intermediate materials". Section 4.3.2 lists all the relevant information that a manufacturer of non-plastic intermediate materials should provide throughout the supply chain.

TKPV technical information sheets 1 to 4 have been revised on this basis. The information sheets already cover all aspects of the Union Guidance on Regulation (EU) No 10/2011 with regard to adhesives. Nevertheless, a formal revision taking into account the Union Guidance and all the new regulations and directives was unavoidable. German and English versions of each information sheet are available on the association's website.

The subject of mineral oils in foods is still occupying German legislators, who would like to introduce new regulations for food packaging made of recycled cardboard. The main source of these oils has been identified as the mineral-oil-based printing inks from newspaper printing.

The fourth draft of the "Mineral oil ordinance" produced by the German Federal Ministry of Food and Agriculture for recycled cardboard has been completed. The final changes made after the hearing for industry associations in July 2017 will not be published until the notification procedure has started.

The draft states that food contact materials made from paper, paperboard or cardboard can only be manufactured from recycled fibre and distributed if they have a functional barrier which ensures that no mineral oil aromatic hydrocarbons (MOAH) can be transferred from the food contact materials into the food. A transfer is considered not to have taken place up to a detection limit of 0.5 milligrams of all MOAH per kilogram of food or food simulant. The limits originally proposed for mineral oil aliphatic hydrocarbons were removed, together with the concentration limits in cardboard.

It is not yet clear whether Germany will take independent action in this area or whether the final draft will be handed over to the EU Commission.

The food industry and all the industries that supply food packaging producers are still facing the problem of false-positive results from analyses. Beeswax, vegetable oils and colophonium produce similar MOSH/MOAH results to mineral oil and can be difficult to distinguish from it.

However, in Regulation No 10/2011/EC, materials that contain mineral oil saturated hydrocarbons (MOSH) and/or mineral oil aromatic hydrocarbons (MOAH) are listed and approved for the production of food contact articles made from plastic. Examples are medical white mineral oils, which are frequently used in cosmetics, and liquid paraffin, which is used in organic farming.

The new TKPV technical information sheet 7 "Low molecular weight hydrocarbon compounds in paper and packaging adhesives" provides comprehensive information about the materials used in adhesives which could be detected as MOSH/MOAH in migration analyses but which have been adequately analysed for food packaging. It also provides guidance on selecting suitable raw materials. An English version is currently being prepared.

At the end of 2017 Food Federation Germany made available a toolbox that will help to prevent the migration of unwanted mineral oil hydrocarbons into food. It provides minimisation methods for all the stages of food manufacturing and for the packaging materials. TKPV expects the chapter on adhesives to be amended during the next revision and has already submitted its proposed text.

Adhesives in paper recycling:
Another important and ongoing aspect of TKPV's work involves assessing the impact of adhesive applications on paper recycling. The current calls by the EU Commission and Germany's Federal Environment Ministry for higher recycling quotas for paper and the end of exports of "poor quality" waste paper to China will require new efforts to be made to improve the recycling process.

In this context, TKPV is taking part in a detailed dialogue with the paper industry and scientific institutions. With the support of TKPV, the International Research Group on De-Inking Technology (INGEDE) has developed the testing procedure "INGEDE Method 12 – Assessment of the Recyclability of Printed Paper Products – Testing of the Fragmentation Behaviour of Adhesive Applications", which has been incorporated into all the relevant eco-labels on a national and European level. The method evaluates the removability of an adhesive application from the recycling process using a screen. The method can be used for all applications of non-water-soluble and non-re-emulsifiable adhesives. Scientific studies which would allow all other types of adhesive application to be evaluated with regard to this recycling process have not yet been carried out.

In addition, TKPV has successfully completed a clustering project in cooperation with INGEDE. Adhesive applications no longer need to be tested using INGEDE method 12, if the adhesive has a minimum softening point and the application takes into account the minimum film thickness and the minimum horizontal expansion. The scorecard of the EPRC (European Paper Recycling Council) has already been amended. Following discussions between the TKPV and RAL, the guidelines for awarding Blue Angel certification are also expected to change. Afera (the European Adhesive Tape Association) has arranged for TKPV to be represented on the EPRC.

Cyclos-htp is evaluating the recyclability of packaging on behalf of DSD and will produce a catalogue of its results by the end of 2019. The evaluation covers the use of adhesives. INGEDE method 12 has been temporarily replaced as the assessment method by PTS-RH:021/97 (October 2012). At the end of 2019 this will be supplemented or replaced by a new PTS method which is better suited to the evaluation of adhesives in the recycling process for packaging.

REACH/CLP:
Another key area of TKPV's work involves activities relating to the new European Chemicals Regulation "REACH" (Registration, Evaluation, Authorisation and Restriction of Chemicals).

Under the provisions of REACH, it is necessary to provide substance data and, in particular, exposure information to ensure that substances are handled safely. For this reason, downstream users of substances must inform the registrants about their use. In order to guarantee that this communication takes place, ECHA has developed a "Use Descriptor" model. On the basis of this model, TKPV has developed use scenarios for the production and the common applications of paper and packaging adhesives. The use scenarios have been included in the FEICA use maps for adhesive production and application.

The concentration limits for labelling as a hazardous substance under the current provisions of the CLP Regulation for methylisothiazolinone as an in-can preservative will in future be changed to correspond with the existing limits for chloromethylisothiazolinone/methylisothiazolinone (3:1). Because products that require no labelling will then fall below the limit for methylisothiazolinone, many producers of aqueous polymer dispersions that are used as a raw material for water-based dispersion adhesives are in a transition phase. This is presenting them with problems because of the lack of resources for replacement products.

Standards activities:
TKPV has been successful in ensuring that German manufacturers of paper and packaging adhesives have expert representation on the Adhesives for Paper, Cardboard, Packaging and Disposable Sanitary Products work group of the European standardisation committee CEN/TC 193 and the German mirror committee DIN/NMP 458. For ten years, the IVK Technical Board has been monitoring a number of DIN, CEN and ISO standards committees and mirror committees, including those referred to above. At the TKPV meeting on 13 June 2019 in Cologne, TKPV decided to bring an end to its own monitoring activities and to obtain information about any changes from the Technical Board.

Networking:
IVK has taken out a subscription with the Federal Institute for Risk Assessment (BfR) on behalf of TKPV to ensure that TKPV is informed directly about changes to the recommendations for food contact materials. In addition, at TKPV's request IVK is now a member of Food Federation Germany, which gives it rapid access to a wide range of information sources. TKPV has a permanent representative on the Food Federation Germany working group for food contact products.

FEICA:
Due to their importance at a European level, the topics of adhesives for materials and articles intended to come into contact with food, MIGRESIVES, FACET, GMP and REACH continue, at the request of TKPV, to be discussed extensively with the European Paper and Packaging and FACET working groups as well as the REACH working group at FEICA. The aim is to establish a joint position for the European adhesives industry and to find solutions to these important issues. TWG PP has also investigated mineral oils in foods and has produced technical information

sheets on the subject. The working group is currently cooperating with associations from the raw materials industry to investigate a variety of raw materials used in hot melts for the packaging industry in order to determine the migration potential of mineral oil hydrocarbons.

Other topics dealt with:
- The work of the Postpress Technical Committee of the Research Institute for Media Technologies (FOGRA)
- The research forum of the PTS
- The work of the Industry Association for Food Technology and Packaging (IVLV) in relation to food packaging
- Microplastics

Technical Committee
Footwear and Leather Adhesives (TKS)

The Technical Committee Shoe Adhesives (TKS) coordinates public relations activities on behalf of German manufacturers of adhesives for footwear materials. In addition, the committee supports German and international standardisation activities and acts as the point of contact for technical evaluations and information specific to certain market segments as part of the work of the German Adhesives Association (IVK) and the German Chemical Industry Association (VCI) with regard to regulatory issues.

Standards activities
The committee focuses on supporting the development of German and European standards to define the basic properties of footwear adhesives.

These activities include standards for:
- Minimum requirements for footwear bonds (requirements and materials)
- Tests to check bond strength in footwear (peel resistance tests)
- Processing (determining the ideal activation conditions and sole positioning tack)
- Durability (colour change caused by migration, thermal resistance of lasting adhesives)

The provision and availability of standardised reference test materials and reference test adhesives are crucial for the ability to perform a large number of tests. The selection and speci-fication of the reference test materials and adhesives are continuously evaluated and updated on the basis of the latest technology. TKS is currently preparing for an audit in this area in 2018/2019.

The TKS information sheet entitled "Problemlösung bei Fehlern in der Schuhherstellung" (Troubleshooting in footwear manufacture) is available to download from IVK's website. It provides assistance and tips for emergencies and service issues that occur during the use of adhesives.

Training

Providing training in adhesive bonding for employees in the footwear industry is another of the tasks that TKS is responsible for. The first courses took place in Zweibrücken in 1990 and in Pirmasens in 1992.

TKS aims to share adhesives knowledge that makes it possible to identify practical problems more quickly, develop solutions and take measures to minimise and even eliminate potential sources of faults in the future.

Against this background, TKS has developed a training concept in conjunction with IFAM/ Bremen and PFI/Pirmasens. It takes the form of a practical seminar called "Angewandte Klebtechnik in der Schuhindustrie" (Applied adhesive technology in the footwear industry). It was held for the first time in November 2005 with more than 20 participants. They gained an in-depth understanding of adhesive technology and the associated issues. The training focuses on providing practical examples to demonstrate best practices in bonding and typical problems. The participants can apply what they have learned in practical exercises, which makes the theoretical information easier to understand for users working in shoe production.

Technical Committee
Structural Adhesives and Sealants (TKSKD)

In the light of the growing number of adhesive applications in which adhesive bonds play a structural role, the Technical Committee Structural Adhesives and Sealants (TKSKD) has been focusing on a range of current technical issues relating to structural adhesives and sealants in order to support the manufacturers of structural adhesives and sealants which are members of the corresponding work group (AKSKD), together with raw material producers and research Institutes involved in this field.

Because structural adhesives are used for a wide variety of applications in many different industries (for example, car, train and aircraft production and shipbuilding, the electrical and electronics industry, the household appliances industry, medical technology, the optical industry, mechanical engineering, plant construction and equipment manufacturing, wind power and solar energy), the committee covers technical subjects of general interest and those relating to specific market segments.

The main areas of the committee's work include:
- **Training in adhesive technology:** Thorough training for users of adhesives and the resulting clear understanding of its applications help to ensure that adhesives are used successfully, correctly and in accordance with the requirements of specific production processes. The training is therefore in the interests of adhesive manufacturers. This is reinforced by the DIN 2304-1 standard, which was published in early 2016. It requires employees who take responsibility for planning and implementing safety-related structural bonds to have the appropriate

qualifications. The committee supports the three-stage DVS/EWF training concept, which has also gained international acceptance, and the preparation of the guidelines "Kleben – aber richtig" ("Adhesive Bonding – the Right Way"), which are available on the IVK website.

- **Promoting research:** TKSKD also sees itself as a bridge between industry and scientific activities that take place outside the industrial world and regularly provides information about topics relevant to structural adhesives as part of planned and approved pre-competitive adhesive research projects. Where a need for research is identified, TKSKD works closely with research bodies to submit applications for new projects. After the projects have been approved, it plays an active role on project monitoring committees.

  For instance, the project run by Braunschweig University of Technology to verify the validity of OIT measurements concerning the thermo-oxidative stability of reactive adhesive systems as a cost-effective and time-saving method for improving the temperature stability of adhesive formulations was supported by individual member companies. The latest status of the project has been reported to AKSKD.

  Currently, TKSKD is working with interested research bodies to draw up a project outline on the subject of the wetting envelope, a process that has been successfully used with coating systems, for example. Transferring the process to adhesives will allow for more accurate predictions of the wetting behaviour of surfaces (bonded parts) when liquids (adhesives) are applied to them and therefore of the development of adhesive forces. As many of the properties of adhesives differ from those of conventional coating materials, the existing methods for determining the required parameters must be tested for their applicability to adhesives and, if necessary, modified. The project also aims to evaluate the general suitability of the process.

- **Standards activities:** The experts from IVK member companies who are members of national and international standards committees report regularly on activities relating to structural adhesives in the following working groups:
  - CEN/TC 193   Adhesives
  - ISO/TC 61/SC 11/WG5   Polymeric adhesives
  - DIN NA 062-10-02 AA   Test methods relevant to structural adhesive technology
  - NA 062-10-03 AA   Adhesive bondings in electronic applications
  - DIN NA 062-10 FBR   Board of the adhesives technical department
  - DIN NA 087-05-06 AA   Adhesive bonding technology in rail vehicle construction
  - DIN/DVS NAS 092-00-28 AA   Adhesive bonding committee
  - DVS AG V 8   Adhesive bonding
  - DVS AG W 4.14   Joining CFRP

- The topics covered over the past few years include:
  - Information on the current status of the internationalisation of Germany's rail vehicle standard DIN 6701 Adhesive bonding of railway vehicles and parts. This standard defines the mandatory requirements placed on companies that manufacture bonded rail vehicles or parts for use on Germany's railways. It is now being turned into a European standard.
  - Information on the work of the standards committee "Quality assurance of adhesive bonds". TKSKD members on this committee have been actively involved in drawing up DIN standard

2304-1 Adhesive bonding technology – *Quality requirements for adhesive bonding processes – Part 1: Adhesive bonding process chain.* The following documents provide more detail on this subject. *DIN SPEC 2305-1 Adhesive bonding technology – Process chain adhesive bonding – Part 1: Advice for manufacturing* explains the requirements for the manufacturing process and DIN SPEC 2305-2 *Adhesive bonding technology – Quality requirements for adhesive bonding processes – Part 2: Adhesive bonding of fibre composite* **materials** describes the special factors involved in bonding fibre composites and the resulting requirements for the bonding process chain. DIN SPEC 2305-3 *Adhesive bonding technology – Quality requirements for adhesive bonding processes – Part 3: Requirements for the adhesive bonding personnel* contains additional information about the requirements for employees in accordance with section 5.2 of DIN 2304-1 for bonds in safety classes S1 to S3. The experts from IVK member companies who sit on the standards committees provide the AKSKD members with regular information about the status of the standards.

- **Chemicals legislation:** As in previous reporting periods, REACH and its requirements for the manufacturers of adhesives and their customers remained a key topic. The member companies are kept regularly informed about current issues. Examples of these include:
  - New inclusions in the list of Substances of Very High Concern (SVHCs).
  - The change in the labelling requirements for methylene diphenyl diisocyanate (MDI) and the current status of the restriction dossier issued by the German Federal Institute for Occupational Safety and Health (BAuA) for diisocyanates in general. The goal of this German initiative is to prevent the introduction of the more far-reaching restrictions on the monomeric diisocyanates that have been requested by some EU member states.
  - The classification of mixtures with organostannic compounds as hazardous materials.
  - The planned process for reporting to poison centres any adhesive formulations that have been classified as harmful to health and the implementation of the process.
  - Testing method for the drinking water approval of anaerobic adhesives: As the guidelines published by the Federal Environmental Agency (UBA) have been withdrawn, there is currently no suitable testing method available for obtaining the necessary drinking water approval for anaerobic adhesives, which are used, among other things, as thread sealants in drinking water systems. An ad-hoc working group within TKSKD has developed practical, standardised guidelines in cooperation with testing institutes and the UBA which permit the use of anaerobic adhesives in drinking water systems.
  - On the initiative of Henkel and Sika Automotive, which are members of the committee, a joint position statement has been drawn up on the subject of the REACH classification of baffles as a product. Baffles are used in lightweight automotive components to reinforce structural parts. They consist of a combination of a structural part and a material that expands when exposed to a high temperature. The shape of the baffle corresponds to the outline of the structural part and it is incorporated into the structural part in the vehicle body shell. In the paint curing oven, the material expands and bonds with the structural part. As the shape of a baffle largely determines its functionality, according to REACH it can be categorised as a product. This means that it does not have to be classified on the basis of the CLP Regulation and no safety data sheets are required.
  - Issues resulting from the requirements of European legislation relating to the circular economy will be of great importance in future.

Public Relations Advisory Board (BeifÖ)

The main task of the Public Relations Advisory Board, which is led by Ulrich Lipper, is to present the German Adhesives Association (IVK) and the key technology of adhesive bonding in all its depth and complexity to the general public in a positive light.

The ongoing public relations work of the German Adhesives Association has proved to be a great success. The subject of adhesive bonding is now firmly established in the print and online media, and the media response is correspondingly high. IVK's press and public relations work has an average annual circulation of around 280 million copies.

The online press platform www.klebstoff-presse.com, which was designed to meet the special requirements of journalists, and IVK's own internet portal www.klebstoffe.com had a consistently high number of page views. Companies, interested end users and journalists can identify the content that is relevant to them at a glance. The association also has a presence on the main social media platforms, such as Facebook, Twitter and YouTube, with "bonding" now taking place on all channels.

The e-paper "Berufsbilder" (Job profiles) has also led to increased web traffic. It highlights for school and university students the varied career opportunities available in the German adhesives industry. Jobs in scientific, technical and commercial fields are explained in a clear, practical and inspirational way. The objective is to encourage the adhesives experts of tomorrow to become familiar today with the industry and the variety of options it has to offer.

The magazine "Kleben fürs Leben" (Bonding for Life) has become a permanent feature of the association's PR activities. Published once a year, it forms an essential component of the communication strategy of the German adhesives industry and plays a key role in promoting the industry's activities. The magazine is free from any form of product or corporate advertising and focuses on further strengthening the positive image of the adhesives industry and documenting the useful features of unique and multifaceted bonding technologies. "Kleben fürs Leben" will be published in 2019 for the eleventh time. Since 2018, readers have also been able to browse through the new online magazine www.kleben-fuers-leben.de. The magazine is distributed to the editorial departments of numerous newspapers and TV and radio stations and to important customer organisations.

The IVK promotional video "Faszination Kleben" (The Fascination of Bonding) demonstrates that adhesives are an essential feature of everyday life in the home, in the construction trades and in industry. It also explains why many technologies of the future and current production processes for everyday objects are only made possible by the use of adhesives. In the space of five minutes, viewers not only discover important facts about the chemical aspects of adhesives and how they function, but also find out about the different areas where adhesives are successfully used. Almost every industry, from the automotive sector to electrical engineering, textiles and clothing, relies on bonding to improve the quality of its products and introduce innovations. Another video entitled "Was unsere Welt zusammenhält" (What Holds Our World Together) has been produced in cooperation with the Association of the European Adhesive and Sealant Industry (FEICA). Both videos are available in German and in English on the IVK website www.klebstoffe.com where they can be viewed or downloaded. They can also be found on the IVK YouTube channel "klebstoffe" (adhesives).

Against the background of the German Federal Government's digitisation strategy, the association has worked together with the German Chemical Industry Fund to develop digital educational materials. A range of interactive infographics designed specifically for use on whiteboards, PCs and tablets in schools has been created to explain the subject of adhesives to school students using modern teaching methods. In addition, the association has commissioned the Institute of the Didactics of Chemistry at the University of Frankfurt to develop a teacher training programme on the subject of "Adhesives and bonding in teaching" in order to give a more in-depth insight into the chemistry of adhesives and to highlight links with the school curriculum.

The new material was completed in time for the start of the 2018/2019 academic year and is available free of charge to teachers throughout Germany.

IVK invites representatives from the business and industry press and daily newspapers to attend its annual press event, which highlights the economic growth of Germany's adhesives industry.

Management

The employees of the German Adhesives Association's office are responsible for coordinating, handling and following up on the wide variety of tasks which come from various committees. The office keeps members up to date on new topics that are important for the industry and its specialised committees. It also serves as an information exchange and a point of contact for the association's members with regard to technical and relevant legal information relating to health and safety at work, environmental and consumer protection and to legislation covering areas such as competition, chemicals, the environment and food.

The management of IVK perceives itself as the representative and expert partner of the adhesives industry. In this role, the management represents the technical and economic interests of the industry among German, European and international authorities. In addition, the employees of the association maintain a close and proactive dialogue with customer, trade and consumer associations, system partners, scientific institutions and the general public. Through active involvement in the advisory boards of major trade shows and trade journals, as well as the working committees of various German and European ministries, the management uses its knowledge and expertise to monitor and support adhesives-related topics and projects on all levels and throughout the entire adhesives value chain.

This also applies to research. In its role as a member of the advisory board of the ProcessNet Adhesive Technology division and the joint committee on adhesive bonding (GAK), the management helps with coordinating publicly funded scientific research projects in the field of adhesives technology.

As part of a broad portfolio of publications, discussions with interested groups, lectures and teaching assignments, the management effectively communicates the adhesives industry's broad range of services and extensive potential for innovation as well as the exemplary commitment of its members to protecting workers, consumers and the environment.

PERSONAL DATA

Honorary Membership & Honorary Presidency

During the 2012 annual general meeting, Arnd Picker was made an Honorary Member and at the same time appointed Honorary President of the German Adhesive Association. By doing so, the association and its members recognised Arnd Picker's achievements in successfully positioning IVK during his 16-year tenure of office as President of its Executive Board. The German Adhesives Association is the world's largest and leading association dedicated to adhesive bonding technology and with an extensive service portfolio for its members.

Honorary Members
Arnd Picker, Honorary President
Dr. Johannes Dahs
Dr. Hannes Frank
Dr. Rainer Vogel
Heinz Zoller

Achievement Medal of the German Adhesives Industry

The German Adhesives Association awards the Achievement Medal of the German Adhesives Industry to individuals for their outstanding service to the adhesives industry and adhesives technology.

The medal was awarded to

Peter Rambusch – May 2018
for the significant contribution he has made to breaking down the organisational barriers between manufacturers of adhesives and producers of self-adhesive tapes which existed until the 1990s. On his initiative, commercial and technical representatives of German adhesive tape manufacturers were admitted into the German Adhesives Association in 1995. Peter Rambusch has continued to represent the interests of the adhesive tape industry on the Executive Board of the German Adhesives Association in a competent, effective and committed way up to the present day. He has not only played an outstanding role in promoting cooperation between two closely related industries, but at the same time has also significantly broadened the skills profile of the German Adhesives Association with regard to all technical adhesive issues.

Marlene Doobe – June 2017
for her decades of commitment to the German adhesives industry. As editor-in chief of the magazine adhäsion KLEBEN ı DICHTEN, she has successfully led this important trade magazine for the adhesives industry for more than 20 years, with professionalism, the highest level of personal dedication and a great deal of passion. During this time and in close cooperation with the German -Adhesives Association, she has developed this magazine into an interdisciplinary 360° communication platform for adhesives technology. Marlene Doobe has made an extremely valuable contribution towards promoting close cooperation between research and industry in the field of adhesives technology and has advanced and shaped the adhesives industry in decisive dimensions.

Dr. Manfred Dollhausen - May 2010
in recognition of his successful involvement in standardisation, with which he made a major contribution to documenting and representing adhesive technology "made in Germany" in national, European and international standards. In addition to that, Dr. Dollhausen recognised early on the invaluable significance of technical cooperation between the adhesives industry and the raw material industry in the 1960s, actively forcing cooperation and thus laying the foundation for the system partnership of both industries that is still successful to this day.

Dr. Hannes Frank – September 2007
in recognition of his many years of dedicated service to the German adhesives industry. As a member of the Technical Committee, he promoted and helped shape both adhesive technology and the image of the adhesives industry. That includes in particular his commitment to small and medium-sized enterprises and their potential to innovate, which is essential for technical and economic development. Dr. Frank is also regarded as a successful pioneer in the field of polyurethane adhesive technology. Furthermore, he has promoted an industry-wide strategy on communication and training, and thus helped to establish adhesives as a key technology of the 21st century.

Prof. Dr. Otto-Diedrich Hennemann - May 2007
in recognition of his scientific work, with which he is largely responsible for promoting and shaping the "bonding system". This includes his research into the -durability of bonds and the implementation of appropriate simulation processes in the auto-motive and aerospace industries. His approach to research was to always focus on concrete applications and the development of added value for system partners.

Committees of the
German Adhesives Association (IVK)

Executive Board

Chairman: Dr. Boris Tasche	Henkel AG & Co. KGaA D-40191 Düsseldorf
Deputy Chairman: Dr. Joachim Schulz	EUKALIN Spezial-Klebstoff Fabrik GmbH D-52249 Eschweiler

Additional Members:

Dirk Brandenburger	Sika Automotive Hamburg GmbH D-22525 Hamburg
Stephan Frischmuth	tesa SE D-22848 Norderstedt
Ansgar van Halteren	Industrieverband Klebstoffe e.V. D-40219 Düsseldorf
Dr. Achim Hübener	Kleiberit Klebstoffe Klebchemie M. G. Becker GmbH & Co. KG D-76356 Weingarten
Dr. Georg Kinzelmann	Henkel AG & Co. KGaA D-40191 Düsseldorf
Patrick Kivits	H.B. Fuller Europe GmbH CH-8006 Zürich
Klaus Kullmann	Jowat SE D-32709 Detmold
Olaf Memmen	Bostik GmbH D-33825 Borgholzhausen
Dr. Bernhard Momper	Celanese Services Germany GmbH D-65844 Sulzbach (Taunus)

Dr. Annett Linemann	H.B. Fuller Deutschland GmbH D-21310 Lüneburg
Dr. Hartwig Lohse	Klebtechnik Dr. Hartwig Lohse e. K. D-25524 Itzehoe
Dr. Michael Nitsche	Bostik GmbH D-33825 Borgholzhausen
Matthias Pfeiffer	Türmerleim GmbH D-67014 Ludwigshafen
Arno Prumbach	EUKALIN Spezial-Klebstoff Fabrik GmbH D-52249 Eschweiler
Dr. Karsten Seitz	tesa SE D-22848 Norderstedt
Dr. Christian Terfloth	Jowat SE D-32709 Detmold
Dr. Christoph Thiebes	Covestro Deutschland AG D-51368 Leverkusen
Dr. Axel Weiss	BASF SE D-67056 Ludwigshafen

Technical Committee Building Adhesives

Chairman: Dr. Norbert Arnold	UZIN UTZ Aktiengesellschaft D-89030 Ulm

Additional Members:

Dr. Thomas Brokamp	Bona GmbH Deutschland D-65531 Limburg
Dr. Michael Erberich	WULFF GmbH & Co. KG D-49504 Lotte

Manfred Friedrich	Sika Deutschland GmbH D-48720 Rosendahl
Dr. Frank Gahlmann	Stauf Klebstoffwerk GmbH D-57234 Wilnsdorf
Stefan Großmann	Sopro Bauchemie GmbH D-65203 Wiesbaden
Holger Hartmann	Celanese Services Germany GmbH D-65844 Sulzbach (Taunus)
Dr. Matthias Hirsch	Kiesel Bauchemie GmbH u. Co. KG D-73708 Esslingen
Michael Illing	Forbo Eurocol Deutschland GmbH D-99028 Erfurt
Frank Mende	Bostik GmbH 33829 Borgholzhausen
Dr. Maximilian Rüllmann	BASF SE D-67056 Ludwigshafen
Dr. Martin Schäfer	Wakol GmbH D-66935 Pirmasens
Dr. Jörg Sieksmeier	ARDEX GmbH D-58430 Witten
Hartmut Urbath	PCI Augsburg GmbH D-40589 Düsseldorf
Dr. Steffen Wunderlich	Kleiberit Klebstoffe Klebchemie M. G. Becker GmbH & Co. KG D-76356 Weingarten

Technical Committee Wood and Furniture Adhesives

Vorsitzende: Daniela Hardt	Celanese Services Germany GmbH D-65844 Sulzbach (Taunus)

Additional Members:

Wolfgang Arndt	Covestro Deutschland AG D-51368 Leverkusen
Holger Brandt	Follmann & Co. GmbH & Co. KG D-32423 Minden
Christoph Funke	Jowat SE D-32709 Detmold
Oliver Hartz	BASF SE D-67056 Ludwigshafen
Dr. Thomas Kotre	Planatol GmbH D-83101 Rohrdorf-Thansau
Jürgen Lotz	Henkel AG & Co. KGaA Standort Bopfingen D-73438 Bopfingen
Dr. Marcel Ruppert	Wacker Chemie AG D-84479 Burghausen
Dipl.-Ing. Martin Sauerland	H.B. Fuller Deutschland GmbH D-31582 Nienburg
Holger Scherrenbacher	Kleiberit Klebstoffe Klebchemie M. G. Becker GmbH & Co. KG D-76353 Weingarten

Technical Committee DIY and Consumer Adhesives

| Chairman: Dr. Dirk Lamm | tesa SE D-22848 Norderstedt |

Additional Members:

Frank Avemaria	3M Deutschland GmbH D-41453 Neuss
Seda Gellings	UHU GmbH & Co. KG D-77815 Bühl
Dr. Nils Hellwig	Henkel AG & Co. KGaA D-40191 Düsseldorf
Dr. Florian Kopp	RUDERER KLEBETECHNIK GMBH NL-85600 Zorneding
Ulrich Lipper	Cyberbond Europe GmbH A H.B. Fuller Company D-31515 Wunstorf
Henning Voß	WEICON GmbH & Co. KG D-48157 Münster

Technical Committee Adhesive Tapes

| Chairman: Dr. Karsten Seitz | tesa SE D-22848 Norderstedt |

Additional Members:

Dr. Achim Böhme	3M Deutschland GmbH D-41453 Neuss
Dr. Thomas Christ	BASF SE D-67056 Ludwigshafen
Dr. Ruben Friedland	Lohmann GmbH & Co. KG D-56567 Neuwied

Dr. Thomas Hanhörster	Sika Automotive GmbH D-22525 Hamburg
Prof. Dr. Andreas Hartwig	IFAM Fraunhofer-Institut für Fertigungstechnik und Angewandte Materialforschung D-28359 Bremen
Lutz Jacob	RJ Consulting D-87527 Altstaedten
Melanie Lack	H.B. Fuller Deutschland GmbH D-21335 Lüneburg
Dr. Thorsten Meier	certoplast Technische Klebebänder GmbH D-42285 Wuppertal
Dr. Ralf Rönisch	COROPLAST Fritz Müller GmbH & Co. KG D-42203 Wuppertal
Dr. Jürgen K. L. Schneider	TSRC (Lux.) Corporation S.a.r.l. L-1930 Luxemburg
Michael Schürmann	Henkel AG & Co. KGaA D-40191 Düsseldorf

Technical Committee Paper and Packaging Adhesives

Chairman: Arno Prumbach	EUKALIN Spezial-Klebstoff Fabrik GmbH D-52249 Eschweiler

Additional Members:

Dr. Elke Andresen	Bostik GmbH D-33829 Borgholzhausen
Dr. Olga Dulachyk	Gludan (Deutschland) GmbH D-21514 Büchen
Dr. Gerhard Kögler	Wacker Chemie AG D-84479 Burghausen

Dr. Anja Köth	tesa SE D-22848 Norderstedt
Dr. Thomas Kotre	Planatol GmbH D-83099 Rohrdorf-Thansau
Volker Liebich	Beardow Adams GmbH D-60386 Frankfurt
Dr. Bernhard Momper	Celanese Services Germany GmbH D-65843 Sulzbach
Dr. Werner Praß	Türmerleim GmbH D-67014 Ludwigshafen
Dr. Peter Preishuber-Pfluegl	BASF SE D-67056 Ludwigshafen
Alexandra Roß	H.B. Fuller Deutschland GmbH D-31566 Nienburg
Dr. Christian Schmidt	Jowat SE D-32709 Detmold
Julia Szincsak	Follmann Chemie GmbH D-32423 Minden
Dr. Monika Toenniessen	Henkel AG & Co. KGaA D-40191 Düsseldorf

Technical Committee Footwear and Leather Adhesives

| Chairman:
Dr. Rainer Buchholz | RENIA Ges. mbH chemische Fabrik
51076 Köln |

Additional Members:

| Wolfgang Arndt | Covestro Deutschland AG
D-51368 Leverkusen |

Martin Breiner	Kömmerling Chemische Fabrik GmbH 66929 Pirmasens
Andreas Ecker	H.B. FULLER Austria GmbH A-4600 Wels
Dr. Martin Schneider	ARLANXEO Deutschland GmbH D-41538 Dormagen

Technical Committee Structural Adhesives and Sealants

| Chairman: Dr. Hartwig Lohse | Klebtechnik Dr. Hartwig Lohse e. K. D-25524 Itzehoe |

Additional Members:

Dr. Beate Baumbach	Covestro Deutschland AG D-51368 Leverkusen
Jürgen Fritz	Evonik Resource Efficiency GmbH D-79618 Rheinfelden
Ralf Fuhrmann	Kömmerling Chemische Fabrik GmbH D-66929 Pirmasens
Dr. Oliver Glosch	Weiss Chemie + Technik GmbH & Co. KG D-35703 Haiger
Dr. Stefan Kreiling	Henkel AG & Co. KGaA Standort Heidelberg D-69112 Heidelberg
Dr. Erik Meiß	IFAM Fraunhofer-Institut für Fertigungstechnik und Angewandte Materialforschung D-28359 Bremen
Bernhard Schuck	Bostik GmbH D-33829 Borgholzhausen
Frank Steegmanns	Stockmeier Urethanes GmbH & Co. KG D-32657 Lemgo

Julius Weirauch	3M Deutschland GmbH D-41453 Neuss
Artur Zanotti	Sika Deutschland GmbH D-72574 Bad Urach

Advisory Board for Closed Substance Cycle Waste Management

Dr. Jörg Dietrich	POLY-CHEM GmbH D-06766 Bitterfeld-Wolfen
Dr. Ruben Friedland	Lohmann GmbH & Co. KG 56567 Neuwied
Elena Gaida	BASF SE 67056 Ludwigshafen
Jürgern Germann	3M Deutschland GmbH 41453 Neuss
Daniela Huch	Cyberbond Europe GmbH A H.B. Fuller Company 31515 Wunstorf
Dr. Peter Krüger	Covestro Deutschlad AG 51373 Leverkusen
Michael Lang	tesa SE 22848 Norderstedt
Linn Mehnert	Wacker Chemie AG 84489 Burghausen
Dr. Stefan Mundt	Henkel AG & Co. KGaA 40589 Düsseldorf

Timm Schulze	Jowat SE 32758 Detmold
Dr. Axel Weiss	BASF SE 67056 Ludwigshafen

Public Relations Advisory Board

Sprecher: Ulrich Lipper	Cyberbond Europe GmbH A H.B. Fuller Company D-31515 Wunstorf

Additional Members:

Rolf J. Blaas	Dow Deutschland Anlagengesellschaft mbH D-65824 Schwalbach
Ansgar van Halteren	Industrieverband Klebstoffe e. V. D-40219 Düsseldorf
Sebastian Hinz	Henkel AG & Co. KGaA D-40191 Düsseldorf
Daniela Huch	Cyberbond Europe GmbH A H.B. Fuller Company 31515 Wunstorf
Oliver Jüntgen	Henkel AG & Co. KGaA D-40191 Düsseldorf
Timm Koepchen	EUKALIN Spezial-Klebstoff Fabrik GmbH D-52249 Eschweiler
Thorsten Krimphove	WEICON GmbH & Co. KG D-48157 Münster
Jens Ruderer	RUDERER KLEBETECHNIK GMBH D-85600 Zorneding
Dr. Christine Wagner	Wacker Chemie AG D-84479 Burghausen

Working Group Building Adhesives

Chairman:
Dr. Rüdiger Oberste-Padtberg

ARDEX GmbH
D-58430 Witten

Working Group Wood and Furniture Adhesives

Chairman:
Klaus Kullmann

Jowat SE
D-32758 Detmold

Working Group Industrial Adhesives

Chairman:
Dr. Boris Tasche

Henkel AG & Co. KGaA
D-40191 Düsseldorf

Working Group Adhesive Tapes

Chairman:
Dr. René Rambusch

certoplast Technische Klebebänder GmbH
D-42285 Wuppertal

Working Group Paper and Packaging Adhesives

Chairman:
Dr. Thomas Pfeiffer

Türmerleim GmbH
D-67014 Ludwigshafen

Working Group Raw Materials

Chairman:
Dr. Bernhard Momper

Celanese Services Germany GmbH
D-65844 Sulzbach (Taunus)

Working Group Structural Adhesives and Sealants

Chairman:
Dirk Brandenburger

Sika Automotive Hamburg GmbH
D-22525 Hamburg

Working Group DIY and Consumer Adhesives

Chairman:
Dr. Dirk Lamm

tesa SE
D-22848 Norderstedt

Working Group Foam Adhesives

Chairman:
Falk Potthast

Jowat SE
D-32758 Detmold

Working Group Footwear and Leather Adhesives

Chairman:
Dr. Rainer Buchholz

RENIA Ges. mbH chemische Fabrik
D-51076 Köln

Management

Ansgar van Halteren	Senior Executive
Dr. Axel Heßland	Managing Director, "Technology & Environment"
Klaus Winkels	Managing Director , "Law"
Dr. Vera Haye	Executive Assistant to Management
Michaela Szkudlarek	Assistant to Senior Executive
Danuta Dworaczek	"Technology & Environment"
Martina Weinberg	Conferences and Conventions
Natascha Zapolowski	"Technology & Environment"
Elke Wegrich	"Internet, Social Media, Communication" (on parental leave)

Honorary President

Arnd Picker

Rommerskirchen

Honorary Members

Dr. Johannes Dahs	Königswinter
Dr. Hannes Frank	Detmold
Arnd Picker	Rommerskirchen
Dr. Rainer Vogel	Langenfeld
Dipl.-Chem. Heinz Zoller	Pirmasens

Holders of the Medal of the German Adhesives Industry

Peter Rambusch	Wuppertal
Marlene Doobe	Eltville
Dr. Manfred Dollhausen	Odenthal
Dr. Hannes Frank	Detmold
Prof. Dr. Otto-D. Hennemann	Osterholz-Scharmbeck

REPORT 2018

FCIO – Austria

FCIO – Austria

The Austrian Association for Flooring adhesives was founded 2008, as a successor from the Austrian Adhesives Association, which was dissolved in 2007. We are a selfstanding part of the Austrian Chemical Industry Federation and the Federal Economic Chamber of Austria.

The Austrian Association of Flooring Adhesives has at the moment 9 members.

Mission statement and services

FCIO-Berufsgruppe Bauklebstoffe (Flooring Adhesive Group) with its 9 member companies, is a legal based association within the Chemical Industry Federation, which represents and defends the interests of the Flooring adhesive Industry in Austria. As an official body, we are mandated by law to give input to all legal and economic issues, which are relating to our business. We stay in permanent contact with our Authorities and our Trade Unions and have representatives in steering committees of various scientific institutions and in working groups of different Austrian Ministries as well as standardization bodies. We help our members to comply with all the demanding laws and regulations in the field of Safety, Health and Environment, especially REACH and CLP or VOC's by preparing various guidance papers and give them legal advice in labor relations.

Together with our clients we are also running education programs for apprentices and trainees.

Organisation and structure

We are part of the Austrian Chemical Industrie Federation FCIO and have our offices in Vienna, Austria

President: Bernhard Mucherl/Murexin

Director: Klaus Schaubmayr

Fachverband Klebstoff-Industrie Schweiz

REPORT 2018

FKS – Switzerland

FKS – Switzerland

Professional Association of the Swiss Adhesive and Sealant Industry

Mission Statement
The association supports its members with regard to adhesive and sealant production, in particular through:
* Representing the interests of the Swiss adhesive and sealant industry with regard to public authorities and associations, inclusive participation in legislative tasks
* Participation in professional expert panels to strengthen the cooperation with public authorities and national and international associations
* Statistics and base information about the Swiss adhesive and sealant market, which provides Swiss and European public authorities with the basic information needed to support their decision-making processes
* Technical clarification and expert reports to promote the customers confidence in the members of the association
* Regular exchanges of information and experience among the members in order to develop the quality of the products
* Organisation of specialised expert lectures

Market developments, directives and measures
Measures for environmental protection and safety in production, packaging, transport, application and disposal are based on the monitoring of market developments and the existing directives.

The measures help to ensure that the services provided always meet the highest requirements of the market.

Member of FEICA
(Association of the European Adhesive and Sealant Industry)
The association is a member of FEICA, which represents the interests of its national associations at the international level in relation to cooperation with international organisations.
FEICA regularly provides the national associations with information concerning developments in Europe.

Service Profile
* Sales statistics about turnover and volume in Switzerland from member companies
* National standards
* Technical clarification and expert reports
* Exchange of information and experience
* Information about developments of regulations
* Annual meeting of members in spring and autumn
* Access to FKS information by the use of member log in on the internet

Organisation & Structure
- President
  Heinz Leibundgut
- Vice-President
  Marcel Leder-Maeder

Members:

Alfa Klebstoffe AG
APM Technica AG
Artimelt AG
ASTORtec AG
BFH Architektur, Holz und Bau
Collano AG
Distona AG
Dow Europe GmbH
Emerell AG
EMS CHEMIE AG
ETH Zürich
H.B. Fuller Europe GmbH

Henkel & Cie. AG
Jowat Swiss AG
Kisling AG
merz+benteli ag
nolax AG
Pontacol AG
Sika Schweiz AG
Türmerleim AG
Uzin Utz Schweiz AG
Wakol GmbH
ZHAW – Zurich University of Applied Sciences

Contact Information

President	Vice-President	Secretariat
Heinz Leibundgut Uzin Utz Schweiz AG Ennetbürgerstrasse 47 6374 Buochs Phone: +41 (0) 41 922 21 31 Fax: +41 (0) 41 922 21 35 Email: heinz.leibundgut@ uzin-utz.com Email: info@fks.ch	Marcel Leder-Maeder Türmerleim AG Hauptstrasse 15 4102 Binningen Phone: +41 (0) 61 271 21 66 Fax: +41 (0) 61 271 21 74 Email: marcel.leder@ tuermerleim.ch	Fachverband Klebstoff- Industrie Schweiz Silvia Fasel Bahnhofplatz 2a 5400 Baden Phone: +41 (0) 56 221 51 00 Fax: +41 (0) 56 221 51 41 www.fks.ch

vereniging lijmen en kitten

REPORT 2018

VLK – Netherland

VLK – Netherland

VLK

Industrial Association

The Vereniging Lijmen en Kitten (VLK) represents the adhesives and sealants industry in the Netherlands. Because of its technical and economic interests, it is an important player on the European market.

The adhesives and sealants industry in the Netherlands focuses primarily on business to business sales. Adhesives and sealants are mainly used in industry and in the construction sector. The VLK estimates that around 75% of the adhesives and sealants the industry produces are sold in the Netherlands.

For the member companies the VLK is:
- *A point of contact* for the general public and public bodies, including inspection and regulatory organisations, institutions of civil society, consumers and other players
- *A spokesperson* for the industry supporting practical legislation on a European level
- *A source of information* and a helpdesk for legislation about materials, such as REACH and the CLP Regulation, and for construction issues, including the CPR and CE marking
- *An organisation that protects* the image of adhesives and sealants and of the adhesives and sealants industry
- *A facilitator* for knowledge sharing and networking.

The VLK is a member of FEICA (the Association of the European Adhesive & Sealant Industry) which represents the industry on a European level.

Services
What can members of the VLK expect?

• Lobbying
The VLK works to ensure that legislation and regulations that affect the development, production and sale of adhesives and sealants in the Netherlands are practical and realistic. The industrial organisation represents the interests of the sector through its contacts with the general public and the institutions of civil society. Many new developments are the results of European legislation, which is why the VLK is a member of FEICA.

• Networking
The VLK is at the centre of a network of companies and plays an important role in bringing them together. Its activities go beyond the adhesives and sealants industry. The VLK maintains contacts with other companies in the supply chain and with academic institutions.

• Knowledge sharing
The VLK identifies relevant information and makes it available to its members via the website and its online newsletters. Member companies also have access to specially developed tutorials on the VLK website. The VLK is made up of three departments and two technical working groups where member companies can share information about developments in legislation and regulations, health and safety, the environment and standards.

• Insights into changes in the market
The flooring adhesives and tile adhesives departments have set benchmarks for developments in the markets that they are involved in. Member companies receive a quarterly report about the most recent activities in the market.

• Helpdesk
The VLK has a helpdesk for issues relating to legislation and regulations that have an impact on adhesives and sealants. Examples include questions about REACH, the CLP Regulation and CE marking. The helpdesk covers both European and Dutch legislation and regulations.

• Positive image and trust
The coordinating role played by the VLK makes it the ideal body to provide information about the adhesives and sealants industry. It helps to ensure the conformity of the industry.

Members
The member companies are the most important feature of the VLK. The industrial department is made up of directors, managers and experts who work on behalf of the VLK central office. The contact details of all the member companies are available on the website at www.vlk.nu/leden. The VLK is not involved in the commercial activities of individual companies.

Organisation
The members of the management board of the VLK are as follows:
• Wybren de Zwart (Saba Dinxperlo BV) – chairman
• Rob de Kruijff (Sika Nederland BV)
• Gertjan van Dinther (Soudal BV)
• Dirk Breeuwer (Forbo Eurocol)

The VLK consists of the following departments:
• Floor adhesives and levelling compounds
• Tile adhesives
• Sealants

The VLK consists of the following working groups:
• Committee for hazardous materials
• Technical committee for sealants
• Technical committee for tile adhesives

Office

In just the same way as paints and printing inks, adhesives and sealants are mixtures. The VLK works closely with the Dutch association for the paints and printing inks industry (VVVF). It has an office within the VVVF and takes part in the association's industry-wide working groups and meetings.

More information

If you need more information, you can contact the VLK via www.vlk.nu.

VLK

Loire 150, 2491 AK, The Hague, The Netherlands
Phone: + 31 70 444 06 80
Email: info@vlk.nu
www.vlk.nu

EMICODE®

Protection from indoor air contaminants

The need to protect people from indoor air contaminants and calls from users for the provision of low-emission products resulted in the foundation of the Association for the Control of Emissions in Products for Flooring Installation, Adhesives and Construction Products (GEV) and the establishment of the EMICODE® labelling system in February 1997. This initiative was supported by leading manufacturers of flooring installation products which are also members of the German Adhesives Association.

The German adhesives industry had originally developed the GISCODE system in the early 1990s in collaboration with the employers' liability insurance organisation for the construction industry, with the aim of helping installers with their choice of flooring installation product. The GISCODE labelling system gives installers a quick overview of products that comply with health and safety legislation.

Under the terms of legislation on hazardous substances, installers of parquet and other types of floor coverings can now only be exposed to high concentrations of volatile solvents in exceptional cases and after taking special protective measures (Technical Rules for Hazardous Sub-

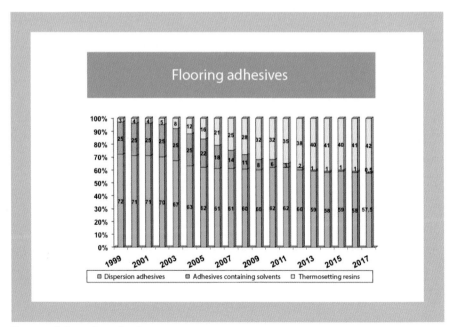

The use of dispersion adhesives

stances (TRGS) 610). The use of substitute substances that represent a much lower risk has become the norm. This has led to a significant number of products containing solvents being phased out over recent years.

Other organic compounds are used in addition to solvents. Some of these are contaminants of raw materials and others are non-volatile compounds. Although they are only present in low concentrations, they can be emitted into the indoor air by the products over long periods.

For this reason, a new generation of solvent-free, very low-emission flooring installation products was developed. These are highly recommended because of their low impact on indoor air quality. GEV now has more than 150 member companies from 21 countries and is steadily growing, particularly in Europe.

In order to provide processors and consumers with reliable guidance in the light of the large number of different measurement processes that are in use, the objective qualification and labelling system called **EMICODE**® was introduced. This makes it possible to evaluate and compare flooring installation materials and other chemical building products on the basis of their emissions. At the same time, it gives manufacturers a strong incentive to continuously improve their products.

The EMICODE® classes are based on a precisely defined test chamber examination and demanding classification criteria. Adhesives, levelling compounds, precoats, substrates, sealants, fast-drying screeds and other building products which are marked with the GEV label EMICODE® EC 1 to indicate that they are "very low in emissions" produce the lowest possible level of indoor air contaminants and odours. Unlike other systems, with EMICODE® the manufacturers themselves are responsible for labelling, while GEV has samples taken of products on the market by independent institutes for monitoring purposes. Another difference is that GEV does not allow for any compromises on quality. Technically questionable environmental criteria are not permitted for reasons of sustainability.

This voluntary initiative is a systematic continuation of the efforts made to protect the health of processors and consumers. EMICODE® provides contracting authorities, architects, planners, tradespeople, building owners and end consumers with transparent and objective guidelines for selecting low-emission flooring installation products. Videos in 12 languages, brochures, tender document templates, technical documents and the organisation's regulations are available on the website www.emicode.com.

By extending EMICODE® to include products such as joint sealants, parquet varnishes and oils, polyurethane foams, floor screeds, reactive coatings and window films etc., GEV has responded to market requirements for the classification of other products that are not traditional floor installation materials and also to requests from the market and from industry to make it possible to differentiate between products on the basis of their environmental impact.

Chairman of the Executive Board	Stefan Neuberger, Pallmann
Chairman of the Technical Advisory Board	Hartmut Urbath, PCI GmbH
Managing Director of GEV	Klaus Winkels, Attorney-at-law

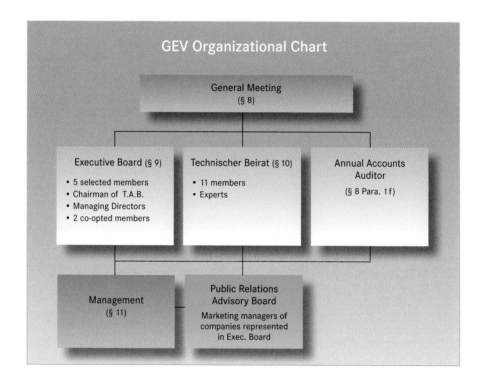

Gemeinschaft Emissionskontrollierte Verlegewerkstoffe,
Klebstoffe und Bauprodukte (GEV)
RWI 4-Haus
Völklinger Straße 4
D-40219 Düsseldorf
Phone +49 (0) 2 11-6 79 31-20
Fax +49 (0) 2 11-6 79 31-33
Email: info@emicode.com
www.emicode.com

FEICA – Fédération Européenne des Industries de Colles et Adhésifs (Association of the European Adhesive and Sealant Industry) – has represented the interests of Europe's adhesives industry since 1972. As the umbrella organisation for 15 national adhesives associations in Europe, it also represents the common interests of 20 individual enterprises, as well as primarily internationally operating adhesive manufacturers and producers of polyurethane foams.

The office of FEICA was operated until the end of 2006 together with the German Adhesives Association with headquarters in Düsseldorf, Germany. Since January 2007, FEICA's headquarters have been located in Brussels.

FEICA's Objectives

With the support of its members, FEICA advocates the common interests of our industry throughout Europe. Moreover, it is also responsible for representing the interests of its members at institutions of the European Union.

External Relations

To obtain information about projects and decrees of European institutions as quickly as possible (European Parliament, European Council, European Commission, Directorate General), FEICA maintains close contact to **CEFIC** (European Chemical Industry Council) and other European associations, of which many from the DUCC group (**D**ownstream **U**ser of **C**hemicals **C**o-ordination) have joined.

Service for IVK Members

As the largest member of FEICA, IVK represents the interests of German adhesive manufacturers in the Executive Board, in the European Technical Board and in various other specialised committees of the European association. This combination assures the members of the German Adhesives Association a corresponding, free-of-charge service as well as all necessary information and regular contact on a European level.

Contact Address

FEICA – Avenue Edmond Van Nieuwenhuyse 2, BE 1160 Brussels www.feica.eu

EUROPEAN LEGISLATION AND REGULATIONS

Relevant Laws and Regulations
for Adhesives

Hazardous Material Law

- **REACH Regulation** – Regulation (EC) No 1907/2006 of the European Parliament and of the Council of 18 December 2006 concerning the Registration, Evaluation, Authorization and Restriction of Chemicals (REACH), establishing a European Chemicals Agency, amending Directive 1999/45/EC and repealing Council Regulation (EEC) No 793/93 and Commission Regulation (EC) No 1488/94 as well as Council Directive 76/769/EEC and Commission Directives 91/155/EEC, 93/67/EEC, 93/105/EEC and 2000/21/EC
- **CLP Regulation** – Regulation (EC) No 1272/2008 of the European Parliament and of the Council of 16 December 2008 on classification, labelling and packaging of substances and mixtures, amending and repealing Directives 67/548/EEC and 199/45/EC and amending Regulation (EC) No 1907/2006
- **Fees Regulation REACH** – Commission Regulation (EC) No 340/2008 of 16 April 2008 on the fees and charges payable to the European Chemicals Agency pursuant to Regulation (EC) No 1907/2006 of the European Parliament and of the Council on the Registration, Evaluation, Authorization and Restriction of Chemicals (REACH)
- **Fees Regulation CLP** – Commission Regulation (EU) No 44/2010 of 21 May 2010 on the fees payable to the European Chemicals Agency pursuant to Regulation (EC) No 1272/2008 of the European Parliament and of the Council on classification, labelling and packaging of substances and mixtures
- **Chemicals test method Regulation** – Council Regulation (EC) No 440/2008 of 30 May 2008 laying down test methods pursuant to Regulation (EC) No 1907/2006 of the European Parliament and of the Council on the Registration, Evaluation, Authorization and Restriction of Chemicals (REACH)
- **Biocidal Products Regulation** – Regulation (EU) No 528/2012 of the European Parliament and of the Council of 22 May 2012 concerning the making available on the market and use of biocidal products
- **PIC Regulation** – Regulation (EC) No 649/2012 of the European Parliament and of the Council concerning the export and import of hazardous chemicals (The new text was adopted on 4 July 2012 and will be applicable from 1 March 2014)
- **POP Regulation** – Regulation (EC) No 850/2004 of the European Parliament and of the Council of 29 April 2004 on persistent organic pollutants and amending Directive 79/117/EEC
- **Chemical Agents Directive (CAD)** – Council Directive 98/24/EC of 7 April 1998 on the protection of the health and safety of workers from the risks related to chemical agents at work (fourteenth individual Directive within the meaning of Article 16(1) of Directive 89/391/EEC).
- **Carcinogens and Mutagens Directive (CMD)** – Directive 2004/37/EC of the European Parliament and of the Council of 29 April 2004 on the protection of workers from the risks related to exposure to carcinogens or mutagens at work (sixth individual Directive within the meaning of Article 16(1) of Council Directive 89/391/EEC).

Circular economy

- **EU action plan for the Circular Economy** – Communication from the Commission to the European Parliament, the Council, the European Economic and Social Committee and the Committee of the Regions – COM(2015) 614 final
- **Ecodesign directive** – Directive 2009/125/EC of the European Parliament and of the Council of 21 October 2009 establishing a framework for the setting of ecodesign requirements for energy-related products (Text with EEA relevance)
- **Plastic Strategy** – Communication from the Commission to the European Parliament, the Council, the European Economic and Social Committee and the Committee of the Regions a European Strategy for Plastics in a Circular Economy – COM(2018) 28 final
- **Plastic Waste:** a European strategy to protect the planet, defend our citizens and empower our industries - Published on: 16/01/2018

Waste Legislation

- **Waste Framework Directive** – Directive 2008/98/EC of the European Parliament and of the Council of 19 November 2008 on waste and repealing certain Directives
- **List of wastes** – Commission Decision 2000/532/EC of 3 May 2000 replacing Decision 94/3/EC establishing a list of wastes pursuant to Article 1(a) of Council Directive 75/442/EEC on waste and Council Decision 94/904/EC establishing a list of hazardous waste pursuant to Article 1(4) of Council Directive 91/689/EEC on hazardous waste
- **Packaging Waste Directive** – European Parliament and Council directive 94/62/EC of 20 December 1994 on packaging and packaging waste

Immission protection

- **Directive on Ambient Air Quality** - Directive 2008/50/EC of the European Parliament and of the Council of 21 May 2008 on ambient air quality an cleaner air for Europe
- **Directive on industrial emissions** – Directive 2010/75/EU of the European Parliament and of the Council of 24 November 2010 on industrial emissions (integrated pollution prevention and control)
- **MCPD Directive** – Directive (EU) 2015/2193 of the European Parliament and of the Council of 25 November 2015 on the limitation of emissions of certain pollutants into the air from medium combustion plants
- **VOC emissions, paints and varnishes** – Directive 2004/42/CE of the European Parliament and of the Council of 21 April 2004 on the limitation of emissions of volatile organic compounds due to the use of organic solvents in certain paints and varnishes and vehicle refinishing products and amending Directive 1999/13/EC
- **Emissions Trading Directive** – Directive 2003/87/EC of the European Parliament and of the Council of 13 October 2003 establishing a scheme for greenhouse gas emission allowance trading within the Community and amending Council Directive 96/61/EC

- **PRTR Regulation** – Regulation (EC) No 166/2006 of the European Parliament and of the Council of 18 January 2006 concerning the establishment of a European Pollutant Release and Transfer Register and amending Council Directives 91/689/EEC and 96/61/EC
- **F-Gases Regulation** – Regulation (EU) No 517/2014 of the European Parliament and of the Council of 16 April 2014 on fluorinated greenhouse gases and repealing Regulation (EC) No 842/2006

Water Legislation

- **Water Framework Directive** – Directive 2000/60/EC of the European Parliament and of the Council of 23 October 2000 establishing a framework for Community action in the field of water policy
- **Dangerous Substances Directive** – Directive 2006/11/EC of the European Parliament and of the Council of 15 February 2006 on pollution caused by certain dangerous substances discharged into the aquatic environment of the Community

Hazardous Materials Transportation Law

- Council Directive 96/49/EC of 23 July 1996 on the harmonization of the laws of the Member States with regard to the transport of dangerous goods by rail
- Council Directive 94/55/EC of 21 November 1994 on the harmonization of the laws of the Member States with regard to the transport of dangerous goods by road
- Council Directive 95/50/EC of 6 October 1995 on uniform procedures for checks on the transport of dangerous goods by road
- Directive 98/91/EC of the European Parliament and of the Council of 14 December 1998 relating to motor vehicles and their trailers intended for the transport of dangerous goods by road and amending Directive 70/156/EEC relating to the type approval of motor vehicles and their trailers
- Regulation (EC) No 2099/2002 of the European Parliament and of the Council of 5 November 2002 establishing a Committee on Safe Seas and the Prevention of Pollution from Ships (COSS) and amending the Regulations on maritime safety and the prevention of pollution from ships
- Council Directive 96/35/EC of 3 June 1996 on the appointment and vocational qualification of safety advisers for the transport of dangerous goods by road, rail and inland waterways
- Directive 2008/68/EC of the European Parliament and of the Council of 24 September 2008 on the inland transport of dangerous goods

International Treaties
GHS – Globally Harmonized System of Classification and Labelling of Chemicals

STATISTICS

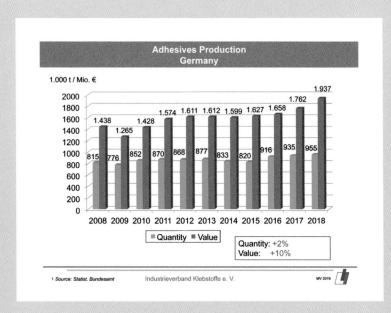

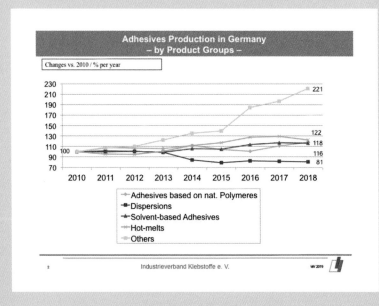

The German Adhesives Industry 2019
– Heterogeneous Environment –

Geopolitical Risks
- Escalation of trade disputes
- Further debt trends
- Decision-making in the EU
- Brexit-Chaos

Slow down of industrial production
- Moderate increase in global GDP
- Decreasing global IPX
- Negative IPX for Germany
- Weakening in the sectors of automotive and electronics

Exchange Rates
- Slight tailwind coupled with more stable dollar

Raw Materials
- Supply situation in general is stable

→ Continuing risks increasingly clouding the customer industries using adhesives

3   Industrieverband Klebstoffe e. V.   MV 2019

Growth forecast continues to decline due to ongoing economic and geopolitical risks

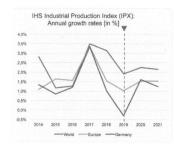

IHS Industrial Production Index (IPX): Annual growth rates [in %]

- **Growth forecast** for 2019 again lowered for all leading countries.
- **Slower growth** in the Euro-Zone for 2019 due to ongoing political uncertainties and restrained export business.
- **Growth of German industry** influenced by external factors such as trade uncertainties, Brexit complications and export weakness.

Source: IHS World Economic Service Mai 2019

4   Industrieverband Klebstoffe e. V.   MV 2019

The German Adhesives Industry
– Development of selected Customer Industries in Germany –

	Share in % of total	2017	Prognosis 2018	Prognosis 2019
Manufacturing	**100**	**2,9**	**0,6**	**0,4**
Transportation	24,1	2,4	-0,5	-1,0
Food, Beverage & Tobacco	9,2	1,3	0,2	0,8
Paper / Printed Matters	3,0	0,9	-1,0	-0,8
Metals & Metal Products	12,7	4,0	0,1	-0,8
Plant & Machinery	13,1	3,8	2,2	2,4
Electrical & Optical Equipment	10,8	5,3	2,0	1,5
Chemical Industry	7,3	1,4	-1,4	0,1
Woodworking (without Furniture)	1,3	4,2	1,2	0,9
Construction Industry	--	**2,3**	**1,0**	**4,3**

Source: IHS World Industry Service Mai 2019

5   Industrieverband Klebstoffe e. V.   MV 2019

The German Adhesives Industry
– Market Sentiment Survey –

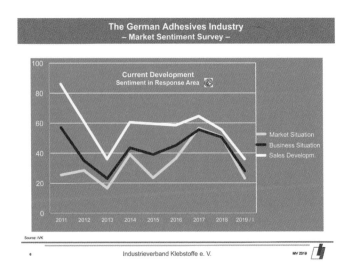

Source: IVK

6   Industrieverband Klebstoffe e. V.   MV 2019

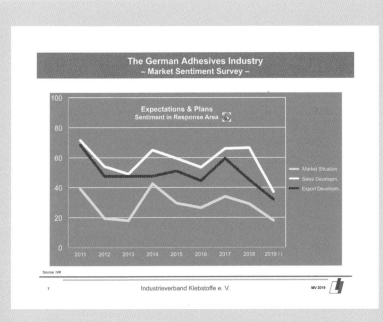

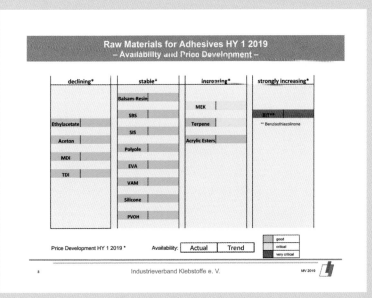

The German Adhesives Industry 2019
– Summary of economic Development –

➢ Economic indicators once again corrected downwards confirming slowing growth trend

➢ Chemical Industry "weakens" ⇨ early-warning indicator

➢ Exchange rates and raw material situation tend to stabilize economic situation

➢ Sentiment in the German adhesives industry tends to be less optimistic

Source: DIW Mai 2019, ifo, ifw, IWH, IVK

9   Industrieverband Klebstoffe e. V.   MV 2019

STANDARDS

Adhesives standards

Early in the 1950s, adhesives standardisation started in Germany at a national level. This initiative continued with the issuing of a terminology standard and other standards for various fields (wood, shoes, flooring adhesives). DIN standards issued during that period by the German Institute for Standardisation (DIN) still apply on a domestic level, unless they have been repealed or replaced by European standards. On 23 January 2013, DIN formed the new committee "Adhesives; Test Procedures and Requirements", which will in future function as an umbrella committee for the DIN standards committees that deal with adhesives. The reason is that most of the standards required are in the meantime available in the form of DIN EN standards, thus making national standards committees dormant. This had the consequence that, in those cases in which a revision was decided on a European level, no active mirror committee and therefore no possibilities for participation were available. The formation of the committee "Adhesives; Test Procedures and Requirements" thus represents a milestone in the reorganisation of adhesives committees at DIN. It is now possible to spin off new subordinate committees relatively quickly and therefore to react more quickly to new requirements than was previously the case. Therefore and due to the integration of other DIN committees dealing with adhesives, the NMP (Standards Committee Materials Testing) sees itself as being well equipped for the future.

In 1961, the "COMITÉ EUROPÉEN DE NORMALISATION" (European Committee for Standardization, CEN) was founded. Thereafter, CEN established various technical committees (Technical Committee 52 "Safety of toys", Technical Committee 67 "Adhesives for tiles", Technical Committee 193 "Adhesives", Technical Committee 261 "Packaging" and Technical Committee 264 "Indoor air quality") to establish and revise standards in the field of adhesives. Thus for the first time the opportunity was created to develop a broad set of standards applicable throughout Europe, and which was specifically aimed at adhesives manufacturers and consumers. This brought with it the chance for them to better represent their own economic and technical interests. Agreements which CEN signed with the "INTERNATIONAL ORGANIZATION FOR STANDARDIZATION" (ISO) meant that the global ISO standards were accepted as European standards and the two organisations went on to jointly work out and issue additional standards.

European standards are generally publicised in the three official languages of the European Union, i.e., German, English and French, and conform with one another from a technical perspective. European standardisation supports trade across the EU by reducing trade barriers and is part of the framework of efforts to achieve technical harmonisation. European standardisation is supported by the German Adhesives Association. Unlike other standards, e.g., international ISO standards, European standards (EN) are binding at a European level. Both the European Court of Justice and national courts within the EU are obliged to base their jurisdiction on European standards.

Specialist testing of adhesives and bonded joints is an important prerequisite for the success of the bonding process. The German Adhesives Association (IVK) has developed a online tool, which is available at www.klebstoffe.com, to provide users with support in this area.

The tool is a database that for the first time summarises all the relevant standards, directives and other important guidelines at a glance. The list currently consists of more than 800 individual documents. It includes not only the relevant German, European and international standards (DIN, EN, ISO), but also standards that were not specifically created for use in the field of bonding, but have become well established in this area.

All the documents are categorised on the basis of the type of adhesive, the application and the contents of the document. Practical search functions allow users to find the subject areas that are relevant to them. A brief description of the contents of each document helps users to determine whether it answers their questions or meets their specific testing requirements. If the document is available free of charge, there is a link to the original version. If payment is required, a link is provided to the website where the document can be purchased.

The standards database is kept constantly updated by IVK, in collaboration with DIN Software GmbH and the consultancy company KLEBTECHNIK Dr. Hartwig Lohse e. K. This ensures that users can always identify which testing methods for adhesives and bonded joints meet their needs. http://www.klebstoffe.com/start-normentabelle.html

SOURCES

- ▶ Raw Materials
- ▶ Adhesives by Types
- ▶ Sealants
- ▶ Adhesives by Key Market Segments
- ▶ Equipment
- ▶ Technical Consultancy
- ▶ Contract Manufacturing and Filling Services
- ▶ Research and Development

Raw Materials
3M
Alberdingk Boley
ARLANXEO
Avebe Adhesives
BASF
BCD Chemie
Biesterfeld Spezialchemie
Bodo Möller
Brenntag
BYK
Chemische Fabrik Budenheim
CHT Germany
CnP Polymer
Coim
Collall
Collano
Covestro
CTA GmbH
DKSH GmbH
EMS-Chemie
Evonik
IMCD
KANEKA Belgium
Keyser & Mackay
Krahn Chemie
LANXESS
Nordmann, Rassmann
Möller Chemie
Münzing
Omya Hamburg GmbH
ORGANIK KIMYA
PCC Specialties
Perstorp Service
Poly-Chem
Rütgers
Schill+Seilacher
SKZ - KFE
Sonderhoff
Synthomer Deutschland GmbH
Synthopol Chemie
Ter Hell
versalis S.p.A.

Wacker Chemie
Willers, Engel
Worlée-Chemie

Adhesives by Types
Hot-melt Adhesives
3M
Adtracon
ALFA Klebstoffe AG
ARDEX
Artimelt
BCD Chemie
Beardow Adams
Biesterfeld Spezialchemie
Bilgram Chemie
Bodo Möller
Bostik
BÜHNEN
BYLA
CHT Germany
Collano
Dupont
Drei Bond
Eluid Adhesive
EMS-Chemie
EUKALIN
Evonik
Fenos AG
Follmann
Forbo Eurocol
GLUDAN
H.B. Fuller
Fritz Häcker
Henkel
Jowat
Kleiberit
Kömmerling
L&L Products Europe
Paramelt
PCC Specialties
Planatol
Poly-Chem

PRHO-CHEM
Rampf
Rhenocoll
Ruderer Klebtechnik
SABA Dinxperlo
Sika Automotive
Sika Deutschland
SKZ - KFE
Tremco illbruck
TSRC (Lux.) Corporation
Türmerleim
versalis S.p.A.
VITO Irmen
Weiss Chemie + Technik
Zelu

Reactive Adhesives
3M
Adtracon
ARDEX
BCD Chemie
Berger-Seidle Siegeltechnik
Biesterfeld Spezialchemie
Bona
Bodo Möller
Bostlk
BÜHNEN
BYLA
Chemetall
COIM Deutschland
Collano
Cyberbond
DEKA
DELO
Drei Bond
Dupont
Dymax Europe
Fenos AG
fischerwerke
Forbo Eurocol
Gößl + Pfaff
H.B. Fuller
Henkel
Jowat
Kiesel Bauchemie

Kisling Deutschland GmbH
Kleiberit
Kömmerling
L&L Products Europe
LORD
LOOP
LUGATO CHEMIE
merz+benteli
Otto-Chemie
Panacol-Elosol
Paramelt
PCI
Planatol
Rampf
Ramsauer GmbH
Ruderer Klebtechnik
SABA Dinxperlo
Schlüter
SCIGRIP
Schomburg
Sika Automotive
Sika Deutschland
SKZ - KFE
Sonderhoff
STAUF
Stockmeier
Synthopol Chemie
Tremco illbruck
Unitech
Uzin Tyro
Uzin Utz
Vinavil
Wakol
Weicon
Weiss Chemie + Technik
Wöllner
ZELU CHEMIE

Dispersion Adhesives
3M
ALFA Klebstoffe AG
BCD Chemie
Beardow Adams
Berger-Seidle Siegeltechnik
Biesterfeld Spezialchemie

Bilgram Chemie
Bison International
Bodo Möller
Bona
Bostik
BÜHNEN
CHT Germany
Coim
Collall
Collano
CTA GmbH
DEKA
Drei Bond
Eluid Adhesive
EUKALIN
Fenos AG
fischerwerke
Follmann
Forbo Eurocol
H.B. Fuller
GLUDAN
Grünig KG
Fritz Häcker
Henkel
IMCD
Jowat
Kiesel Bauchemie
Klebstoffwerk COLLODIN
Kleiberit
Kömmerling
LORD
LUGATO CHEMIE
Murexin
ORGANIK KIMYA
Paramelt
PCC Specialties
PCI
Planatol
PRHO-CHEM
Ramsauer GmbH
Renia-Gesellschaft
Rhenocoll
Ruderer Klebtechnik
Schlüter
Schomburg

Sika Automotive
SKZ - KFE
Sopro Bauchemie
STAUF
Synthopol Chemie
Tremco illbruck
Türmerleim
UHU
VITO Irmen
Wakol
Weiss Chemie & Technik
Wulff
ZELU CHEMIE

Vegetable Adhesives,
Dextrin and Starch Adhesives
BCD Chemie
Beardow Adams
Biesterfeld Spezialchemie
Bodo Möller
Collall
Distona AG
Eluid
EUKALIN
Grünig KG
H.B. Fuller
Henkel
Paramelt
Planatol
PRHO-CHEM
Ruderer Klebtechnik
SKZ - KFE
Türmerleim
Wöllner

Animal Glue
H.B. Fuller
Henkel
PRHO-CHEM

Solvent-based Adhesives
3M
Adtracon
BCD Chemie
Berger-Seidle Siegeltechnik

Biesterfeld Spezialchemie
Bilgram Chemie
Bison International
Bodo Möller
Bona
Bostik
CHT Germany
COIM Deutschland
Collall
CTA GmbH
DEKA
Distona AG
Fenos AG
Fermit
fischerwerke
Forbo Eurocol
H.B. Fuller
IMCD
Jowat
Kiesel Bauchemie
Kleiberit
Kömmerling
LANXESS
LORD
Otto-Chemie
Paramelt
PCC Specialties
Planatol
Poly-Chem
Ramsauer GmbH
Renia-Gesellschaft
Rhenocoll
Ruderer Klebtechnik
SABA Dinxperlo
Sika Automotive
STAUF
Synthopol Chemie
Tremco illbruck
TSRC (Lux.) Corporation
UHU
versalis S.p.A.
VITO Irmen
Wakol
Weiss Chemie + Technik
ZELU CHEMIE

Pressure-Sensitive Adhesives
3M
ALFA Klebstoffe AG
BCD Chemie
Beardow Adams
Biesterfeld Spezialchemie
Bostik
BÜHNEN
Collano
CTA GmbH
DEKA
Dymax Europe
Eluid Adhesive
EUKALIN
H.B. Fuller
GLUDAN
Fenos AG
Fritz Häcker
Henkel
IMCD
Kleiberit
L&L Products Europe
LANXESS
ORGANIK KIMYA
Paramelt
PCC Specialties
Planatol
Poly-Chem
PRHO-CHEM
Rhenocoll
Ruderer Klebtechnik

Sealants
ARDEX
Berger-Seidle Siegeltechnik
Bodo Möller
Bison International
Bostik
Botament
CTA GmbH
Drei Bond
EMS-Chemie
Fermit
fischerwerke

Henkel
L&L Products Europe
merz+benteli
Murexin
ORGANIK KIMYA
OTTO-Chemie
Paramelt
PCI
Rampf
Ramsauer GmbH
Ruderer Klebtechnik
Schomburg
SCIGRIP
SKZ – KFE
Sonderhoff
Stockmeier
Synthopol Chemie
Tremco illbruck
Unitech
UHU
Wulff

Adhesives by Key Market Segments

Self Adhesive Tapes
3M
Alberdingk Boley
artimelt
Avebe Adhesives
Bodo Möller
BYK
certoplast Technische Klebebänder
CNP-Polymer
Coroplast
DKSH GmbH
Eluid Adhesive
Fritz Häcker
IMCD
LANXESS
Lohmann
Planatol
Schlüter
Synthopol Chemie
Tesa

Paper/Packaging
3M
Adtracon
Alberdingk Boley
ALFA Klebstoffe AG
artimelt
Avebe Adhesives
BCD Chemie
Beardow Adams
Biesterfeld Spezialchemie
Bilgram Chemie
Bison International
Bodo Möller
Bostik
Brenntag
BÜHNEN
BYK
certoplast Technische Klebebänder
CNP-Polymer
COIM Deutschland
Collano
Coroplast
CTA GmbH
DEKA
DKSH GmbH
Eluid Adhesive
EMS-Chemie
EUKALIN
Evonik
Fenos AG
Follmann
Forbo Eurocol
H.B. Fuller
GLUDAN
Grünig KG
Fritz Häcker
Henkel
IMCD
Jowat
LANXESS
Lohmann
Möller Chemie
MÜNZING
Nordmann, Rassmann
Nynas

Omya Hamburg GmbH
ORGANIK KIMYA
Paramelt
PCC Specialties
Planatol
Poly-Chem
PRHO-CHEM
Rhenocoll
Ruderer Klebtechnik
Sonderhoff
Synthopol Chemie
tesa
TSRC (Lux.) Corporation
Türmerleim
UHU
versalis S.p.A.
Wakol
Weicon
Weiss Chemie + Technik
Wöllner

Bookbinding/
Graphics Industry
3M
ALFA Klebstoffe AG
BCD Chemie
Biesterfeld Spezialchemie
Bodo Möller
Brenntag
BÜHNEN
BYK
CNP -Polymer
Coim
Collall
DKSH GmbH
Eluid Adhesive
EUKALIN
Evonik
H. B. Fuller
Fritz Häcker
Henkel
IMCD
Jowat
LANXESS
Lohmann

Möller Chemie
MÜNZING
Nordmann, Rassmann
Omya Hamburg GmbH
ORGANIK KIMYA
PCC Specialties
Planatol
PRHO-CHEM
Sika Automotive
tesa
TSRC (Lux.) Corporation
Türmerleim
UHU
versalis S.p.A.
Vinavil

Wood/Furniture industry
3M
Adtracon
ALFA Klebstoffe AG
BCD Chemie
Berger-Seidle Siegeltechnik
Biesterfeld Spezialchemie
Bilgram Chemie
Bison International
Bodo Möller
Bostik
Brenntag
BÜHNEN
BYK
BYLA
Chemische Fabrik Budenheim
CNP-Polymer
Collall
Collano
Coroplast
CTA GmbH
Cyberbond
DEKA
DKSH GmbH
Eluid Adhesive
EMS-Chemie
Evonik
Fenos AG
fischerwerke

Follmann	Building and Construction Industry
Gößl + Pfaff	including Floors, Walls and Ceilings
Grünig KG	ARDEX
H.B. Fuller	artimelt
Henkel	BCD Chemie
Jowat	Berger-Seidle Siegeltechnik
KANEKA Belgium	Biesterfeld Spezialchemie
Kisling Deutschland GmbH	Bilgram Chemie
Kleiberit	Bodo Möller
Kömmerling	Bona
LANXESS	Bostik
Lohmann	Botament
Möller Chemie	Brenntag
MÜNZING	BÜHNEN
Nordmann	BYLA
Omya Hamburg GmbH	BYK
ORGANIK KIMYA	certoplast Technische Klebebänder
Otto-Chemie	Chemische Fabrik Budenheim
Panacol-Elosol	CnP Polymer
PCC Specialties	Collano
Rampf	Coroplast
Ramsauer GmbH	CTA GmbH
Rhenocoll	DEKA
Ruderer Klebtechnik	DELO
SABA Dinxperlo	DKSH GmbH
Sika Automotive	Emerell
SKZ - KFE	EMS-Chemie
STAUF	Evonik
Stockmeier	Fenos AG
Synthopol Chemie	Fermit
tesa	fischerwerke
Tremco illbruck	Forbo Eurocol
TSRC (Lux.) Corporation	Gößl + Pfaff
Türmerleim	H. B. Fuller
versalis S.p.A.	GLUDAN
Vinavil	Henkel
VITO Irmen	IMCD
Wakol	Kiesel Bauchemie
Weicon	Kleiberit
Weiss Chemie + Technik	Kömmerling
Wöllner	Lohmann
ZELU CHEMIE	LUGATO CHEMIE
	Mapei
	Möller Chemie
	Murexin

MÜNZING
Nordmann, Rassmann
ORGANIK KIMYA
Otto-Chemie
Paramelt
PCC Specialties
PCI
Planatol
Poly-Chem
Rampf
Ramsauer GmbH
Rhenocoll
Schlüter
Schomburg
SCIGRIP
Sika Automotive
Sika Deutschland
SKZ - KFE
Sopro Bauchemie
STAUF
Synthopol Chemie
tesa
Tremco illbruck
TSRC (Lux.) Corporation
Uzin Tyro
Uzin Utz
Vinavil
Wakol
Weicon
Weiss Chemie + Technik
Wöllner
Wulff

Car and Aircraft Industries
ALFA Klebstoffe AG
APM Technica
Beardow Adams
Bison International
Bodo Möller
Brenntag
BÜHNEN
BYLA
certoplast Technische Klebebänder
Chemetall
Chemische Fabrik Budenheim

CHT Germany
CNP-Polymer
Coroplast
Cyberbond
DEKA
DELO
Drei Bond
Dupont
Dymax Europe
Emerell
EMS-Chemie
Evonik
Fenos AG
Gößl + Pfaff
H.B. Fuller
Henkel
Kleiberit
Kömmerling
L&L Products Europe
Lohmann
LORD
Möller Chemie
MÜNZING
Nordmann, Rassmann
Otto-Chemie
Panacol-Elosol
PCC Specialties
Planatol
Polytec
Rampf
Ramsauer GmbH
Ruderer Klebtechnik
SCIGRIP
Sika Automotive
Sika Deutschland
Sonderhoff
Synthopol Chemie
Tremco illbruck
tesa
TSRC (Lux.) Corporation
VITO Irmen
Wakol
Weicon
Weiss Chemie + Technik
ZELU CHEMIE

Electronics
APM Technica
Bison International
Bodo Möller
Brenntag
BÜHNEN
BYLA
certoplast Technische Klebebänder
Chemetall
CHT Germany
Collano
Coroplast
CTA GmbH
Cyberbond
DELO
DKSH GmbH
Drei Bond
Dymax Europe
Emerell
EMS-Chemie
Evonik
Gößl + Pfaff
H.B. Fuller
Henkel
KANEKA Belgium
Kisling Deutschland GmbH
Kömmerling
L&L Products Europe
Lohmann
LORD
Möller Chemie
MÜNZING
Nordmann, Rassmann
Otto-Chemie
Panacol-Elosol
PCC Specialties
Polytec
Rampf
Ruderer Klebtechnik
SCIGRIP
Sika Automotive
Sika Deutschland
SKZ - KFE
tesa
Tremco illbruck

Unitech
UHU
Weicon
Weiss Chemie + Technik

Sanitary Industry
APM Technica
Bilgram Chemie
H.B. Fuller
GLUDAN
Henkel
Jowat
Kömmerling
LANXESS
Lohmann
Nordmann, Rassmann
Nynas
PCC Specialties
Prho-Chem
Sika Automotive
Türmerleim
Vito Irmen

Assembly, General Industry
BCD Chemie
Biesterfeld Spezialchemie
Bodo Möller
BÜHNEN
BYLA
certoplast Technische Klebebänder
Chemetall
CHT Germany
Coroplast
Cyberbond
DEKA
DELO
Drei Bond
Dupont
Gößl + Pfaff
Henkel
KANEKA Belgium
Kleiberit
Kömmerling
L&L Products Europe
Lohmann

Novamelt
Otto-Chemie
Panacol-Elosol
Paramelt
PCC Specialties
Renia-Gesellschaft
Ruderer Klebtechnik
SABA Dinxperlo
Schomburg
SKZ - KFE
Sonderhoff
Synthopol Chemie
tesa
Weicon

Textile Industry
Adtracon
BCD Chemie
Biesterfeld Spezialchemie
Bodo Möller
Bostik
Brenntag
BÜHNEN
Chemische Fabrik Budenheim
CHT Germany
CNP-Polymer
Collano
DEKA
Emerell
EMS-Chemie
EUKALIN
Evonik
H.B. Fuller
Henkel
Jowat
Kleiberit
LANXESS
Möller Chemie
MÜNZING
Nordmann, Rassmann
Nynas
Omya Hamburg GmbH
PCC Specialties
SABA Dinxperlo
Sika Automotive

Synthopol Chemie
tesa
Vito Irmen
Wakol
Wulff
Zelu

Self Adhesive Tapes, Labels
artimelt
BCD Chemie
Biesterfeld Spezialchemie
Bodo Möller
Bostik
Brenntag
CNP-Polymer
Coim
Collano
EMS-Chemie
EUKALIN
Fenos AG
H.B. Fuller
Henkel
IMCD
Jowat
KANEKA Belgium
LANXESS
Möller Chemie
MÜNZING
Nordmann, Rassmann
Nynas
ORGANIK KIMYA
Paramelt
Novamelt
PCC Specialties
Planatol
PRHO-CHEM
Stauf
Sika Automotive
SKZ - KFE
Synthopol Chemie
TSRC (Lux.) Corporation
Türmerleim
versalis S.p.A.
Vito Irmen

Household, Hobby,
Offices, Stationery
Bodo Möller
Bison International
certoplast Technische Klebebänder
CNP-Polymer
Collall
Coroplast
CTA GmbH
Cyberbond
EMS-Chemie
Fenos AG
Fermit
fischerwerke
GLUDAN
Henkel
KANEKA Belgium
LUGATO Chemie
Möller Chemie
Nordmann, Rassmann
Nynas
Omya Hamburg GmbH
Panacol-Elosol
PCC Specialties
Rampf
Ramsauer GmbH
Renia-Gesellschaft
Rhenocoll
SCIGRIP
tesa
Tremco illbruck
TSRC (Lux.) Corporation
UHU
versalis S.p.A.
Weicon
Weiss Chemie + Technik

Footwear & Leather Industry
Adtracon
BÜHNEN
Cyberbond
H.B. Fuller
Henkel
Kömmerling
Renia-Gesellschaft

Ruderer Klebtechnik
Sika Automotive
Wakol
Zelu

Equipment
for Adhesive Handling, Mixing, Dosing
and Application

Baumer hhs
BÜHNEN
Drei Bond
Hardo
Hilger u. Kern
Innotech
IST Metz
Nordson
Plasmatreat
Reinhardt-Technik
Robatech
Rocholl
Scheugenpflug
Sonderhoff
Sulzer Mixpac
Viscotec
Walther

Technical Consultancy
ChemQuest Europe INC.
Hinterwaldner Consulting
Klebtechnik Dr. Hartwig Lohse

Contract Manufacturing and Filling Services
LOOP

Research and Development
IFAM
BFH
ZHAW